Dignified Dying

A Guide

This book is dedicated to:

Pieter Admiraal (1929-2013)

Dutch anesthesiologist and right-to-die pioneer
Royal decoration: *Officer Oranje Nassau*
Janet Good Memorial Award from the Hemlock Society
Honorary member of Right-to-Die NL and
Deutsche Gemeinschaft für humanes Sterben

Dignified Dying

a Guide

Death at Your Bidding

Boudewijn Chabot MD PhD

© B. Chabot MD PhD, Amsterdam 2014
Published by the author
c/o Postbox 3930, 1001 as Amsterdam, Netherlands

first edition: August 2014
second edition: September 2014
third revised an renamed edtion: July 2015
published as e-book July 2015
published as print on demand book July 2015
books can be ordered through www.dignifieddying.com

isbn/ean 978-90-816194-6-2
nur 863

Book design by Gerrit Vroon, Arnhem
Printed by CPI/Koninklijke Wöhrmann, Zutphen

Cover image: Rafael Croonen, Gerrit Vroon

Disclaimer

It is the author's conviction that information about how to achieve a self-chosen, humane death in the presence of one's loved ones should be available to people who have a well-considered and lasting wish to die. In this book, he provides this information based on his experience, on expert pharmacological and toxicological advice, and on reports from relatives who were present.

The author of this book does not in any way wish to encourage forms of suicide that are chosen by depressed people and that are impulsive, lonely and violent. Before undertaking a self-chosen death, someone near the end of life with a well-considered and persistent wish to die should receive professional therapy, palliative care, spiritual comfort if desired, and other help to make life bearable.

There are three methods to hasten death in a well-considered and responsible way: by medication, by helium gas and by voluntarily stopping eating and drinking. These are not suicide methods, but an exercise of the individual's right to die in connection with his or her loved ones. From this book, it will become clear that a humane, self-chosen death requires many time-consuming preparatory steps, which are not compatible with acting on impulse or in a violent way.

It remains the reader's responsibility to comply with all the laws of his or her country and/ or state regarding the topics covered here and other end-of-life decisions. The author is not responsible for failure or any unfortunate outcome.

Table of Contents

Chapter 4 The search for the holy grail

Foreword by Faye Girsh

Past president of the World Federation of Right to Die Societies.*

HELPFUL ADDITION TO THE END-OF-LIFE GENRE

As someone who's been in the right to die field for many years I still find it confusing about what to advise people about the best way to have a peaceful death when help from a doctor is unobtainable, as it is for most people around the world.

Chabot's new book I found to be a further clarification of the major methods with more scientific evidence, clarification of some of the more dubious methods that people talk about, and a clear description of what is needed. He explores what the criteria are for deciding on which method is workable and compares information from several sources.

Chabot presents an excellent analysis of the need for information for those who want a self-chosen, humane death (a good term he employs) outside the law when a doctor will not help or when the person wants to retain control. He looks into the future when the number of old people and patients with long-term chronic disease and impending dementia will engulf the medical profession with their needs to die peacefully.

Taking it out of the hands of already reluctant doctors and empowering individuals, families, and professional end-of-life counselors with this information is the wave of the future, as Derek Humphry did with his pioneering, *Final Exit*. Chabot's book, though quite accessible to the lay reader, offers more scientific data and embeds the information in an historical and somewhat philosophical context. That there is a flurry of these do-it-yourself guides indicates people's hunger for a choice of workable methods for self-empowerment. It is too bad that so many of these techniques still require underground, often illegal, access.

* Five star review on Amazon, September 21, 2014 to the previous, second edition of this book.

Foreword by Libby Wilson MD

Past convenor of the Scottish Friends At The End (FATE)*

Chabot sets out clearly and compassionately those methods of ending one's life effectively and with the least possible distress. This book also has chapters on methods to be avoided because they are not always successful and can be associated with unpleasant side effects before death supervenes.

The use of drug overdosage and of inert gasses like helium or nitrogen are described in detail which can be understood by any reasonably intelligent adult.

Voluntary refusal of food and fluid is also a recommended possibility for the elderly or terminally people which Chabot described in *Taking Control of Your Death by Stopping Eating and Drinking*.

Chabot has the advantage of practicing in The Netherlands which is known for its liberal law on Physician Assisted Suicide, but there is still a demand for knowledge about methods which do not demand medical intervention. In countries like the UK where suicide is legal but 'assisting' is not, this book makes an important contribution which should be available to all those who wish to know how to end a life whose suffering has become intolerable.

* Review to the previous, second edition of this book in FATE Newsletter August 20, 2014.

Foreword by Gerrit Kimsma MD MPh

Dutch family physician and philosopher*

How can you die a self-chosen, dignified death if you face unbearable suffering and, together with your loved ones, you can no longer see a way out? For over 20 years, Dr Chabot has been carrying out research into how people have actually been doing this in the Netherlands over the years, by stopping eating and drinking and by taking lethal drugs. To everyone's surprise, both methods are used much more often than had been thought.

What he discovered is special for two reasons. First, doctors frequently remain out of the picture in such cases, because the act takes place outside the medical domain; and second, because a great deal of consultation and coordination is needed in order to make this a positive experience together with family and friends, in spite of the grief and the other feelings surrounding a farewell. In many countries, including the Netherlands, criminal law constitutes a barrier by criminalizing the act of taking control over one's death in consultation with one's family.

Twelve years after the Dutch Act of 2002, it is clear that physician-assisted dying is failing to meet the desire within society to be able to take control of ending one's life. This book is a powerful plea for do-it-yourself methods to have a place alongside physician-assisted dying. These methods are of particular importance for chronic psychiatric patients, elderly people who wish to forestall their deterioration through Alzheimer's, and elderly people who consider their lives to be 'complete'. These are significant groups in society, and this book makes it very clear to them how they can achieve a dignified end, together with their loved-ones. This book may prove to be of support to them.

*– a past member (1998-2010) of the Review Committee where cases of
  physician-assisted dying are reported;
 – a consultant since 1997 for doctors who plan to perform physician-assisted
  dying and trainer of other consultants;
 – physician-philosopher at the Department of Bioethics, Radboud University
  Nijmegen.

Preface

Two personal experiences

My interest in methods for ending one's life surrounded by loved ones is rooted in two experiences, both of which drew me to the subject fifty years ago. The first was the death of my father, who had a heart condition and who had reached an impasse in his work and love relationship. Before giving me a chance to understand his wish to die, he let himself be taken by a heart attack. 'A quick death', was the cardiologist's verdict; but I knew that it was a hidden suicide. They never appear in the statistics, all those people who silently find a way to bring an end to their mortal suffering 'by accident'.

Second, when I was studying medicine, I was always fascinated by the question: 'Why do people resign themselves to the power of doctors? As a doctor, will I leave people in ignorance and continue to treat them with chemo- or radiation therapy? Even when I know death will probably come in one or two months and the quality of this last stretch of their life will be badly affected by the treatment?' I felt a growing desire to tell people that they had the option of refusing further treatment in order to protect their quality of life when the end was in sight, whether due to illness or to very old age.

A turning point in my life came in 1991 when, for the first and last time, I assisted a woman in a suicide with 9 gram of secobarbital and reported this to the police. This 50 year old woman who did not suffer from any somatic disease had lost her two sons under dramatic circumstances. The Dutch Supreme Court found me guilty without imposing any punishment.[1]* The notorious Chabot case sparked a heated public debate. Some argued that I had been seduced to death while others concluded that this woman had suffered the slings and arrows of outrageous misfortune.[2] For this woman, living on without her sons would have meant the loss of integrity that is the hallmark of suffering.[3]

Since then my life has become focused on researching non-doctor

* From here on, note indicators in the text refer to the notes on p. 117

assisted dignified dying by medication or by voluntary stopping eating and drinking in the Dutch population. The surprising results of this survey study were published in the professional literature.[4] Since then I have elaborated the findings into practical steps that are understandable and workable for a laypersons in different countries.[5]

This guide to dignified dying is certainly not meant as a substitute for physician-assisted dying. However, even in a tolerant societies like Belgium or The Netherlands doctors cannot fulfill all demands for a gentle death. The doctor-centered and autonomy-centered routes to a humane death should not be seen as mutually exclusive alternatives, but rather as complementary ones. Only together can they provide an answer to the demand that a gentle death in the presence of family or friends will become accessible for those individuals who consider this to be of the utmost importance.

A grim subject?

This book deals with what many consider to be a grim subject: preparing for your own death, or that of your loved ones. Talking about it, arranging how it will be done, and perhaps actually doing it: these remain awkward issues. However, there is a growing desire on the part of people nearing the end of their lives to take control of their own death. I am convinced this desire is pervasive.

In most countries the medical and legal professions have joined forces to make it almost impossible to die a gentle death at the time of one's choosing. Notable exceptions in Europe are Belgium, The Netherlands and Switzerland. In Canada the law passed in Quebec (2014) that allows for physician-assisted dying has drawn worldwide attention. In the United States there are famous exceptions starting with Oregon (1997) followed by a number of other Federal States where physian-assisted dying has become a lawful option. Recently their number is increasing almost every year.

However, in countries and states that still oppose physician-assisted dying in 'terminal patients', as they are often referred to, the dominant political view is that life is precious and should be cherished, not ended in a careless fashion. True as this may be, increasing numbers of citizens all around the world feel that the time of one's death is ultimately a private affair. Not only in the immediate circle of one's family and friends but also in relation to one's own spiritual beliefs.[6]

This topic may well remain controversial for decades to come. Therefore this book is not aimed at medical or other professionals, but at lay people who are interested in knowing which dignified options are available in end-of-life situations due to incurable disease causing unbearable physical and psychological suffering that cannot be eased under conditions they deem tolerable. My aim is to show people that as long as they make the necessary preparations well enough in advance, the instruments for choosing a gentle death will be theirs.

What does this book add to existing publications?

Three high-profile books have already discussed ways of choosing one's own death: Final Exit by Derek Humphry, The Peaceful Pill Handbook by Philip Nitschke & Fiona Stewart, and Five Last Acts – The Exit Path by Chris Docker. My book offers a different perspective on several points.

First, the empirical basis: I spent ten years undertaking a survey in the Dutch population to assess the frequency of "dying a gentle death" accompanied by relatives or friends. How do people manage to do it? Everyone knows it happens, but not how often it occurs. By now, two studies conducted in the Netherlands (population about 16 million) have demonstrated that every year there are at least 600 self-chosen deaths by stopping eating and drinking and 300 by medication attended by relatives.[7] No epidemiological data on these two methods for a dignified self-chosen death are available for any other country in the world.

Second, I had the good fortune to find two leading Dutch pharmacists Paul Lebbink (Pharm.D), Annemieke Horikx (Pharm.D), and biochemist-toxicologist Ed Pennings (Ph.D) willing to share their expertise with me in ongoing discussions. They fully endorsed my attempt to provide information on the options for the medication method with lethal doses. This allowed me to incorporate in chapters 2 and 3 their expertise on the appropriate combinations and dosages for a humane self-chosen death. Their contributions were also indispensable in analyzing attempts to find other autonomous methods (chapter 4).

Third, I take a moral stand. It should be obvious that people who choose to die alone are fully entitled to do so and no one should moralize about their choice. It is equally obvious that very old or very ill persons should not be forced to die alone for fear of the legal consequences to those who are present. This last point has been insufficiently dealt with in right-to-die publications. In the last section of the introduction I will return to this sensitive topic: how can you pass away in your intimate circle while taking precautions that may appease the authorities after your death?

Textbox: changes in the third edition

Dignified Dying – A Guide (www.dignifieddying.com) is a thoroughly revised and retitled third edition of *A Way to Die* that was published in September 2014 at the Chicago World Federation Congress of Right-to-Die Societies.

the introduction has been expanded to include recent jurisprudence of the *European Court of Human Rights*. The **echr** has stated that a fundamental right to privacy and a family life includes the right to decide when and how one chooses to die. From cases in Switzerland, Germany, The Netherlands and Ireland, lessons can be drawn that might protect relatives and close friends who have assisted in a self-chosen death.

chapter 1 on basic information regarding lethal drugs has remained unchanged.

nutech meeting san francisco, June 6-7 2015. Methods for a self-chosen death have been discussed by experts from around the world. The insights I gained from them have been assimilated in this book.

chapter 2 on the medication method discusses effective methods that have been witnessed many times. I have removed morphine, phenobarbirtal and tricyclic antidepressants from chapter 2 to chapter 4 because observational reports from different professionals are scarce.

chapter 3 has been expanded to include nitrogen gas.

chapter 4 has been expanded to include a discussion of other lethal gases and of mechanical and medication methods that are still lacking in observational reports by experts.

The previous edition had a concluding chapter 5 on physician-assisted dying in the Netherlands. Developments around self-chosen death in that country are 'confusing' to put it mildly. I have decided to drop this topic from the present edition and wait whether the fog will clear up when in 2016 the Dutch Supreme Court has passed judgment in the case of Abert Heringa.

Introduction
Good or bad death in death-denying societies

Is that how I want to end my life?

As they get older, many people are troubled by visions of what it might be like to linger on for years, needing constant help, with their friends vanishing one by one, the future shrinking, and tiredness making every visit seem too long. What goes through your mind if you find yourself in a nursing home where even your bedtime is decided for you? What might it be like to suffer from all the woes of old age or a crippling disease and to feel your vitality ebbing away, facing the prospect of having to stay in bed 24/7? Is the overriding feeling then a fear of death, or a desire for death?

Elderly people sometimes wonder whether they might be able to die a gentle death. They think, 'Will I really have to spend years being dependent on other people's care before I'm allowed to die? Will I have to spend days waiting for a visit when I can't read or watch TV? How can I take control of my own death?'

Many people feel strongly that they don't want to end their lives in a nursing home. But neither do they want to put their loved ones through the ordeal of finding them dead by their own hand, in some violent or horrible way. Some manage to find their own path to what they consider to be a "good death": passing away peacefully in their own bed, surrounded by those who are dear to them.

Moving into the driver's seat

These days, one often hears people standing up for the right to die 'with dignity'. These people may feel that they are masters of their destiny. They believe in self-determination and autonomy, and feel these values should apply to the end of life as well. Many countries have constitutions in which these fundamental values are enshrined. But when it comes to dying, societies seem to be divided about whether these values should be applied.

What autonomy means in practice is closely linked to prevailing ideas of what makes a 'good death'. One person might say a good death means living as long as possible, while for someone else, it might mean dying while you are still able to think clearly, or to live independently Enormous differences of opinion exist between those who see a good death as something that happens at a time of your own choosing, and those who believe it means waiting until God calls you to Him. There are other areas of disagreement: can a 'good death' take place at home, or is it only possible in an institution? How much care might your loved ones be able to organize and pay for? Are dependency and suffering meaningful experiences.

Some paint a dark picture of the notion that elderly persons should be able to plan their own deaths, emphasizing that autonomy can be manipulated. They point out that what we define as a personal decision, a free choice, is partly determined by our shared beliefs regarding the purpose of life. Don't we have obligations to our children? Wouldn't accepting the notion of a humane self-chosen death disrupt the delicate social fabric of caring for each other?

Certain changes in society may well lead to more acceptance of the idea that there is a 'time to die'. Take the fact that we are increasingly tending to confine the elderly to nursing homes. In such homes, people lose their former social roles. What's more, institutional care is becoming more and more expensive, and not everyone has the resources to afford the best possible care. In such conditions, ending one's life might come to be considered a reasonable alternative to the prospect of a harrowing illness such as Lou-Gehrig disease or the oblivion of Alzheimer's.

Ever since the days of Socrates and Seneca, public figures have acted out their concept of a dignified death. However, the difference today – a situation that is unprecedented in the history of mankind – is that millions of elderly people are slowly dying in institutions when they are already socially "dead"; that is, no longer able to live an ordinary meaningful life. It is not uncommon for patients to be restrained or sedated for the benefit of the staff. Suppose someone who faces the prospect of having to be admitted to such an institution chooses to end his or her life in a dignified fashion. This might be called an autonomous, or even rational, decision. But is it really a free choice? In the end, it comes down to opting for death as the lesser of two undesirable choices.

Can someone, in a self-directed way and without the help of a doctor, die a good death if life is no longer tolerable for them? Because doctors hold the keys to the medicine cabinet, you are dependent upon them – unless you know the way out. Some very old people wish to choose the time of their death when they have had enough of life. Sometimes people with terminal illnesses – people who have by no means had enough of life – do not want to wait to die as a result of their illness. Many strive for an end that is in harmony with the life that they have lived. The question is often asked: how can I remain in control? After all, isn't the manner in which I die a part of my life? Perhaps the most important part for me and for my loved-ones, who will live on with the example that I want to give them, that it is possible to die a good death. The choice should be in your own hands. The Economist has put it this way: 'Although most Western governments no longer try to dictate how consenting adults have sex, the state still stands in the way of their choices about death. An increasing number of people — and this newspaper — believe that is wrong'.[1]

Shared opinions about good or bad death

Anthropologists and historians have studied notions of what constitutes a good death in a wide variety of Western and non-Western societies.[2] They concluded that ideas regarding what constitutes a good death appear to be virtually universal. Three characteristics emerge.
A good death means:
- dying at the end of a long life,
- at home and surrounded by those dear to you,
- from illness or old age, without violence as in suicides by hanging or by firearms.

A bad death, on the other hand, is one that takes place prematurely or violently. It also means dying alone or surrounded by strangers, as often happens in a hospital or in a nursing home.

These values seem to be anchored in the human condition. They help us to make a distinction between suicides that devastate family and friends and ways of ending one's own life that can lead to a good death, a self-directed, dignified death.

Which methods are available for a good death?

For a good death it is important that at least one person from the intimate circle can be present at the self-chosen deathbed. It remains to be seen whether being present at the death scene is legally safe. This depends on the law of the country or US State and, to a large extent, also on its enforcement by the authorities. My focus in this book will be on two methods for a self-chosen death that are effective in at least 90% of cases and physically safe for those present:

1. combinations of drugs, the medication method for short (ch. 1 and 2)
2. inert gases, of which nitrogen is the rising star and helium the waning one (ch. 3).

Textbox. A warning

From April 2015 onward some helium balloon tanks have been diluted with
20% air, which contains about 4-5% oxygen. The dilution must be indicated on the tanks. When a 80 / 20 mixture of helium and air is inhaled death will probably take hours to come or not at all. As long as there are still tanks around with pure helium gas, one can use them for a self-chosen humane death. Philip Nitschke (**pph** handbook online edition) has claimed that nitrogen gas is equally effective as helium. For more information on nitrogen gas see chapter 3

This book would be incomplete without a discussion of the many and varied other methods for a self-chosen death some of which have been around for some time: poisons and other gases. All are usually executed alone, in secret and unexpected to those who loved the deceased. In chapter 4 I will discuss which ones may be effective. The evidence reported by independent observers is scanty. No rate of successes versus failures for these methods is known. Some of them, like e.g. carbon monoxide, are still in an experimental phase as they are risky for those present.

My decision not to discuss these lonely suicide methods in detail is not based on a moral judgment. Anyone who wants to die alone and take the risk of failure has the right to proceed on his own. However, very old or very ill persons prefer not to die alone but in the presence of someone dear and want to be reassured that the method is effective in almost all cases. Only some drugs and some inert gases will suit them.

Stopping eating and drinking

Completely different from all the methods just mentioned is Voluntarily Stopping Eating and Drinking (VSED) under palliative care. This can be a dignified death but the practical, legal and ethical differences compared to the other two are such that I have chosen to discuss this method in a separate book. Let me mention here a few points the reader should know to decide whether he wants to learn more about this oldest of all methods to hasten death.

Only for very ill or very old people, voluntary refusal of food and fluids may be a peaceful and natural way of dying.[3] The aim is to hasten, under palliative supervision, a death that would have come eventually after months or years. A competent person who is seriously ill, or weak because of old age, and who deliberately refuses to drink (apart from some water for mouth care), will become sleepy within a week and die some days later. Palliative care can make it possible for all those concerned to say their goodbyes with all the intensity of emotion that they feel.

My book about voluntary stopping eating and drinking to hasten death (VSED)[4] explains how and why this need not be a gruesome way out. In the polarized public debate some advocates of physician-assisted dying discredit this self-directed route to death: 'The only legal option is to starve oneself to death – a hideous course that many people take in desperation'.[5] Apparently those commentators are still unaware of the research among hospice nurses in Oregon that showed that almost all of the patients they had cared for who had chosen this route had a dignified death.[6] It is not only for hospice patents that thirst can be made bearable. According the knmg (2015), even for the elderly at home, clinical experience has shown that with good oral care and easily obtainable prescription drugs, this can be a dignified way out.[7]

Some doctors are unwilling to supervise the process as they fear that providing palliative care might be considered in religious circles as assistance in suicide. However, palliation by e.g. giving mouth care and preventing bed sores are not considered assistance in suicide by the authorities, These skills can be learned by any compassionate spouse, child, friend or other caring person. Detailed information on VSED that is available in my book and online[8] should be studied beforehand.

In summary, during the process of VSED a compassionate doctor is helpful but not indispensable. A nurse trained in palliative care who teaches some skills to those who want to care for you is more important. Elderly and terminally ill persons who yearn for a hastened death can stop eating first and then drinking to hasten the death they accept is approaching. Thanks to palliative care dying by VSED has become less difficult than it has been for the elderly and terminally ill persons in the past who did not get meticulous mouth care. It is important to realize that you do not have to express your intention to hasten death in words, except to someone you can trust (see appendix 2).

The right to privacy and a family life

For most elderly and ill people the medication and inert gas methods I have just mentioned are within their reach provided they get some help in the preparatory phase from relatives or close friends. Why would anyone take that risk? Close relatives and friends are bound to the person with a strong wish to die by many emotional threads. A good death for someone very old or very ill means to die connected with the people that care about him or her. This includes their being able to let you go, if and only if you can convince them that this is what you really, truly want.

If very old or very ill persons don't want to die alone how can they pass away within their intimate circle and at the same time try to protect those present against legal proceedings? Of course, there will never be any guarantee against harsh treatment by the police and sleepless nights in jail. But with some precautions an encounter with the law will become less likely, as has become evident in recent jurisprudence of the European Court for Human Rights (ECHR).

This important change came about in 2011 when the ECHR acknowledged in the case of Mr. Haas vs. Switzerland that Article 8 of the European Constitution on the fundamental right to "privacy and a family life", includes the right to decide when and how one chooses to die. Two years later the European Court acknowledged the special position of relatives in a case of suicide in Germany.[9]

In 2015, Albert Heringa has been acquitted by a higher court in the Netherlands from assisting in the suicide of his 99 year old mother whom he had given a lethal cocktail of chloroquine and valium.[10] Giving someone medication for a suicide is forbidden by Dutch penal law but the court took his special position as a son into account as well as his video of some steps in the process that ended in her death.

Another acquittal of assisted suicide in a case of progressive MS attracted public attention in Ireland (see Appendix 3).[11] When a friend tried to arrange a travel for this patient to the Dignitas clinic in Switzerland a travel agent alerted the police to the plan. Shortly thereafter the patient was found dead in a wheelchair having taken a lethal dose of barbiturates from Mexico. The friend who was not present at the death was acquitted of the charge of assisting in a suicide.

In view of these legal developments regarding citizens from several European countries – Switzerland, Germany, the Netherlands and Ireland I want to discuss which lessons can be drawn from these cases that may protect relatives and close friends in the near future who want to be present at the self-chosen death of either a very old or very ill person. I am aware that at first sight my suggestions for relatives who live under more repressive jurisdictions may well sound utopian.

How to protect those present at your death

Lawyer William Simmons has suggested improvements
to the Introduction and Chapter 4 that I have accepted with gratitude.

Assisting in a self-chosen and humane death is considered a crime in most European and English-speaking countries and in others. Diverging interests exist between the person who wants to die in the presence of loved ones, and the interests of society in protecting vulnerable citizens from being encouraged to die. The authorities want to establish, and rightly so, that no one else has helped to perform the very last acts that cause death. To my knowledge this topic has not recieved the attention it deserves in right-to-die publications.

Assistance in the last acts that cause death is known to occur among experts in cases of dying by the helium method or with the "plastic bag with sedatives" method. I do not condemn this kind of help under the difficult circumstances that often occur at the end of life. However, every society has a legitimate stake in the safety of the elderly and terminally ill that should be acknowledged and respected.

The recent court cases in Europe give a clue how the tensions might be lessened between the right of the individual to determine time, place and method of his death on the one hand and the obligation of society to protect vulnerable individuals from being encouraged to die on the other. My advice is far from complete but a first start is better than none.

The authorities usually want to have some evidence, first that relatives and friends have not encouraged a very old or very ill person to proceed with a self-chosen death. Second, they want to be pretty sure that the last acts that caused death were performed by the dying person, not by those in attendance.

Regarding the first precaution my advice is that relatives or close friends should provide evidence not only in writing but also on video that the person who wanted to die had seriously considered potential ways of making life more bearable. At least one such conversation about the reasons to proceed with the self-chosen death should be recorded on video or on a smart phone. This should be a real conversation, not a written statement read aloud. Such a recorded interaction will be more convincing for the authorities than a written statement. However, a written statement, such as in an advance healthcare directive, can also be helpful, especially if repeated over a period of time.

In the preparatory phase, very old or very ill people need help from relatives or friends to obtain some of the medicines or some of the equipment for an inert gas method. However in some jurisdictions such purchases may be enough to bring a charge of assisting suicide. This is unfair as having pentobarbital in one's bedside cabinet or inert gases in tanks does not cause death. With the medicine method, death is caused by drinking the liquid pentobarbital or by sprinkling the lethal powder in a cup of yogurt and eating it with a spoon. When using an inert gas it is the responsibility of the person who wants to die to turn the tap on the tanks and drawing the hood over their face. My second advice is that elderly or very ill people should do this them-selves and that it should be recorded on video.

My hope is that elderly or terminally ill people will learn how they can take responsibility for their own death. They may be motivated to do this in time once they understand that by proving it on video they can protect their loved one's against legal proceedings.

In summary, to satisfy the authorities it is most essential that the dying person takes full responsibility for the acts that cause death. These can be recorded by video or smart phone. Together with the recorded conversations in the preparatory phase, this should be pro-vided in the post-mortem as some evidence that no crime, particularly that of assisting suicide by pushing someone over the edge of the cliff, has been committed. This evidence may not be enough but it is a be-ginning that will be welcomed by impartial judges or by members of a jury.

I am aware that we have a long way to go to achieve a compro-mise between the conflicting interests of the individual and those of society. It will require the courage of conviction of many individuals in the face of possible prosecution before the authorities may come to understand and accept that a self-chosen death in the privacy of one's immediate circle is the fulfillment of a long-standing and deeply felt wish. The road of prosecution and harsh punishment is a dead end. It has driven self-chosen and self-performed dying assisted by relatives and friends underground. The Dutch authorities were sur-prised when my nationwide survey unearthed that a self-chosen death in the intimate circle occurred in a substantial number of deaths.[12] In other countries this has not been researched yet as has been done in the Netherlands. Most governments do not finance these nationwide survey studies. Perhaps, they do not like to find out what is going on in the twilight that surrounds dying at home.

Chapter 1

Lethal drugs: what one should know and do

1.1
Confusing information and legal precautions

It is widely known that many suicide attempts fail. One reason for this is the confusing information in the literature on self-inflicted death. For instance, it is often thought that natural substances, such as poisonous plants, can be used in suicide. People cite the death of Socrates from an extract of hemlock (Lat. *Conium maculatum*). Indeed, Plato romanticized Socrates' death as a gentle death. In fact, the poison in hemlock brings on paralysis while the person is fully conscious, accompanied by diarrhea and sometimes even convulsions. Eventually the paralysis reaches the respiratory muscles, causing slow suffocation. Socrates died a slow death by suffocation.

This is by no means an exception. Attempts at hastening death by using natural poisons (from toadstools, snakes, spiders and others) may sometimes succeed, but will *always* be accompanied by excruciating pain or suffocation (see chapter 4).

The mass media are a source of confusion on humane and effective methods. Time and again, they report on the use of potentially lethal medicines, such as the insulin that was used by the physician Harold Shipman to kill elderly patients. Indeed, if the patient is in a frail condition, heavy doses of insulin can be lethal. Nevertheless, the toxicological literature provides evidence that even in extremely large doses, the lethal effect of insulin on healthy persons is uncertain.

Last but not least, medical specialists are a source of confusion. Information on lethal drugs is not part of their training. A death is considered dignified if one falls into a deep sleep first and only then, when unconscious, one dies from cardiac and/or respiratory arrest. Some doctors give authoritative advice what medication will cause death. They don't tell you that for some time you will be aware of very unpleasant symptoms like cardiac pain or suffocation.

Legal precautions

Once one has settled on the decision to end one's life humanely, one should consider taking legal precautions. One risk factor is being discovered before death has occurred and being subjected to unwanted attempts to save one's life: pumping of the stomach, cardiac resuscitation, artificial respiration and drip-feeding. You can refuse all these in a refusal of treatment statement that should be supplemented with an advance directive. Your name must be clearly visible and the document must be dated and signed(see the example in Appendix 1). Be sure any legal form you use meets the particular legal requirements of your country or state. Ask a right-to-die organization in your country for advice.

In 1990, in the case of Nancy Cruzan, the US Supreme Court stated that every *competent* adult has the constitutional right to refuse any medical treatment.[1] You need to ensure that no one will try to 'save your life' once you lose consciousness and are no longer competent.

This raises two questions. First, how is 'competence' defined? The answer is that if you can understand what alternatives for treatment remain and can clearly express why you are turning them down and choosing your route to death, you are competent whether or not a psychiatric diagnosis has been made in the past.[2] Your reasons for refusing further treatments should be written down in a few sentences, with the date and your signature. If a doctor raises any doubts about your competence, ask a social worker or mental health professional to interview you about your decision to die and make a formal statement of your competence.

The second question is: who can stand up for your rights? You should grant power of attorney beforehand to a legal representative, authorizing this person to make decisions on your behalf regarding refusal of treatment once you have become incompetent. This legal representative must ensure that your advance directive refusing hospital admission and all medical treatment is respected. Usually this is a spouse, child or close friend, who supports your decision and is assertive enough to oppose pressure from ambulance personnel, nursing staff or doctors insisting that your life must be 'saved'.

For elderly persons who know no trusted person to act on their behalf, it may be hard to find representatives, but it is an essential precaution. Volunteers from organizations like Compassion and Choices are often willing to step in.

Why is death the only option left?

It is of the utmost importance to share the reasons for your decision to end your life with your loved ones or close friends, and, if possible, with your family doctor. Why? First, it is important to ensure that this decision is not a mistake and that other possibilities for palliative care have not been overlooked. Second, it is important to have the support of at least one person in order to prevent unwanted life-saving rescue efforts. And third, the dialogue with your loved ones can be valuable in helping them come to terms with your decision and their loss.

Anyone who wants to hasten death by drugs or gas should be aware of the difficult position in which relatives and close friends will find themselves. To be sure, one or more family members and friends may oppose the decision. They may be overwhelmed by anger and grief when confronted with their loved one's wish to die. If you are really determined, you'll have to explain patiently why you feel that your life is not worth living anymore. This may involve an emotional struggle with loved ones. They may feel rejected by your refusal of care, certainly after they have done all they can to help you to continue to live a meaningful life rather than end it. Some withdraw from these discussions, as they are unable to accept your decision and prefer not to be informed about the planned date. Other relatives are less troubled by ambivalence, because they suspect that if they themselves were in the same situation, they would make the same decision.

Sometimes there are no loved ones or close acquaintances whom one can trust. In such situations, it is always possible to speak to an outsider with experience in giving guidance to people with a wish to die. Volunteers from right-to-die organizations in your country will have experience in counseling.[3] Your family doctor should be sensitive to questions about palliative options and may have experience in that new specialty.

1.2
Principles: what one should know

The pharmacist and toxicologist I consulted insist that the reader be introduced to certain principles before turning to chapter 2. Textbox 1.1 gives a summary of these principles.

Textbox 1.1 List of principles that apply to all lethal drugs

– Generic and brand names of drugs
– Drug storage life
– Lethal dose and body weight
– Why sleeping pills (benzodiazepines) are necessary
– The need for detox: stopping current drug use

People often think that the only important thing is to know the name and lethal dose of a drug. However, if one wants to be certain of a painless death, there are other things one needs to know as well. Death by drugs is a painful death when these drugs are not taken in combination with sleeping pills. It is also important to know that to prevent a failed attempt, many drugs require a period of 'detox' (detoxification) to allow the person to become 'clean'. For how long should a person be 'clean' before the planned date? To maximize the chances of success, you should know the answers to this and other questions. This is not the kind of situation in which things should be left to chance.

Generic and brand names
Four types of drugs are suitable for taking control of your death: barbiturates, opiates (strong pain killers), chloroquine (anti-malaria drugs) and tricyclic antidepressants. You should choose which to use, based on how easy it is for you to obtain them and how much confidence you have in them. Each will be discussed in a separate section.

Every drug has a variety of names. One of these is the generic name and refers to the chemical substance it contains. Since this name is the same in every country, this is the name that will be used here. In addition, every drug has different trade or brand names under which it is sold in pharmacies. These brand names will often differ from one country to the next. To help readers, some brand names will be given in parentheses. For instance, diazepam is a generic name, while Valium, Q-pam, Valcaps or Vazepam are all brand names for this chemical substance.

Drug storage life

According to the Dutch pharmacist consulted for this book, Veterinarian Nembutal that is sold in the Netherlands can be stored for three years (this is stated on the packaging). The pharmacists recommend that lethal drugs, whether they are in liquid, tablet or powder form, be kept in sealed bottles in a dark place. In these conditions they can be stored for ten years or sometimes more; that is, way beyond the date that is usually indicated on the package.

Using a refrigerator does not prolong the storage life of lethal drugs. In any case, a refrigerator is a poor storage place, because it carries the risk that others may use the drugs for an impulsive suicide. It is of no use to add moisture-resistant granules.

Lethal dose and body weight

Toxicologists use the term LD-50 as a shorthand for 'Lethal Dose in 50% of cases.' The LD-50 indicates the dose at which a chemical substance is lethal for 50% of the population studied. Toxicologist Ed Pennings stresses that animal data give confusing information regarding the dose that would be lethal in human beings. His advice for the drugs in this book is to at least triple the LD-50 that is given in toxicological textbooks.

The late Dutch anesthesiologist Pieter Admiraal used to say that no one has ever woken up after taking 6 grams of a barbiturate. However, a few people are so-called 'rapid metabolizers' who break down a lethal drug quickly with the help of enzymes in the liver which is almost impossible to determine beforehand. Therefore, there is still a very small chance of failure when an overdose is taken of five times the amount. The only example of which I am aware was a man in Oregon who did not die after taking the ± 10 grams of pentobarbital that had been legally prescribed by a doctor for a physician-assisted death.

The lethal doses given in chapter 2 apply to persons with a body weight of between 60 and 100 kg (132-222 pounds). Anyone who weighs more than 100 kg should increase the lethal dose by 10% for every 10 kg (22 pounds) of extra weight. The lethal dose can be reduced by 10% for every 10 kg under 60 kg.

Why are sleeping pills necessary?

It is not enough to take a lethal drug. Why not? All lethal drugs kill because the heart stops beating or one can no longer breathe, or both. If one were not in a deep sleep, this would cause pain in the chest or suffocation. This is why the group of sleeping pills called benzodiazepines (or BDs) need to be taken as well. Taken by themselves, even in very large doses they may sometimes lead to death, but often fail to do so.

BDs have three important properties. First, taken in a large dose, they induce such a deep sleep that one could not possibly be conscious when the heart or breathing stop, and so one would not experience this. Second, they suppress certain side effects of lethal drugs, such as painful muscle spasms or epileptic seizures. And third, a large dose of a sleeping pill may reinforce the effect of the lethal substance.

There are about twenty different BDs on the market, and most of them will not induce a deep sleep that lasts for 24 hours or more. All lethal drugs, with the exception of barbiturates, need to be accompanied by a long-acting BD that induces such a long sleep. Some lethal drugs act so quickly that they have to be taken with a fast-acting BD as well, which helps one to fall asleep rapidly. Although other sleeping pills may be added, no amount of them can replace a long-acting sleeping pill.

There are only two long-acting BDs that the pharmacists Lebbink and Horikx advice: diazepam or flurazepam (table 1.1). Used in the recommended dose, they induce a deep sleep for about 48 hours, which will start about 30 or 60 minutes after taking the drug. There is not much choice either when it comes to fast-acting BDs: midazolam, temazepam or lorazepam.

In chapter 2, the information about each drug will be summarized in tables. The tables show which drugs need to be combined with a fast-acting BD. Note, for instance, that oxazepam (Ox-Pam, Serax, Zaxopam) does not induce a long-lasting sleep and better should not be used instead of a long-acting benzodiazepine (or BD).

Table 1.1 Long-acting and fast-acting sleeping pills (BDs), with dosage*

Long-acting sleeping pills	dosage	Fast-acting sleeping pills	dosage
flurazepam (Dalmane, Durapam) 15 or 30 mg**	500 mg	**midazolam** (Versed), 7.5 mg or 15 mg tablets	150 mg
diazepam (Valium, Q-pam, Vazepam)	500 mg	**temazepam** in solution (Normison), 10 mg or 20 mg capsules	400 mg
		lorazepam (Alzapam, Ativan, Loraz), 1 mg or 2.5 mg	25 mg

* For all drugs one should take notice of the Tmax, i.e. the time to maximum concentration in the blood. Also the T½, i.e. the time to eliminate half of the medicine is relevant. These values are given in any pharmacological handbook.

** Be aware that medicines come in tablets that contain different dosages. For instance, Flurazepam tablets may contain either 15 mg or 30 mg. One therefore needs at least 34 15 mg-tablets or 17 30 mg-tablets.

Detoxification: stopping existing drugs use

Perhaps you are already using one of the drugs you intend to use for your self-chosen death on a daily basis. For instance, you may be using an opiate to control pain or a benzodiazepine to fall asleep. In such cases, you will have to gradually 'kick' this habit first to become 'clean' (to detoxify). Why?

Some of the drugs used for a dignified death produce 'habituation'. Most people will be familiar with this effect in the case of alcohol: an inexperienced drinker will get drunk more quickly than someone who has been drinking for years. To achieve the same effect, the heavy drinker has to drink more. Regular use also decreases the brain's sensitivity to the drug, so regular users of opiates or sleeping pills usually have to take more and more to achieve the same effect. Opiates, in particular, produce habituation very rapidly, so much so that for a person who takes them on a daily basis, even trebling the dose would not lead to death. One has to stop using opiates for three weeks to become 'clean'. If that would be impossible due to pain, you would be better switching to one of the other lethal medicines discussed in chapter 2.

Table 1.2 Habituation in some lethal drugs and whether it is necessary to stop taking them in advance

	Habituation	Necessary to quit in advance?
Opiates	+++	Yes, definitely
Barbiturates	+	Not necessary, as they are not prescribed any more
Chloroquine	—	No
Antidepressants	—	No
Benzodiazepines	+	Recommended

Chloroquine and antidepressants do not lead to habituation. This means that people who are taking these drugs do not need to stop taking them for a period prior to the self-chosen death. Sleeping pills also lead to habituation, but the effect is less extreme than with opiates. We recommend that you cut down on your use of them in order to achieve the necessary long and deep sleep.

1.3
Preparations: what one should do

Now that we have discussed the principles that apply to all lethal drugs, we turn to the preparatory steps that apply to all of them.

Textbox 1.2 Preparations: list of issues to consider

– How to get hold of lethal drugs
– Can one trust drugs obtained through the Internet?
– How to take the drugs
– Anti-emetics to prevent a failed attempt due to vomiting
– Should drugs be taken on an empty stomach?
– Interaction with other drugs that may hamper the lethal effect
– Alcohol: drink in moderation before taking drugs
– When the person has fallen asleep: information for family and friends

How to get hold of lethal drugs

These days, most people use the Internet to obtain lethal drugs. Others are lucky enough to get them in small quantities from their doctor, often with a story about fake symptoms that the doctor may well see through. Some travel to a non-Western country, where many drugs are sold over the counter without a prescription. Elderly or ill persons who are no longer able to travel can ask a friend or relative to buy them when on holiday and send them through the post.

Barbiturates (discussed in detail in section 2.1) and opiates (section 2.2) are substances that fall under international drug control conventions. There are strict controls on the sale, transport and possession of these drugs. Pharmacies can be prosecuted if they sell these drugs without a prescription. Companies based in non-Western countries do not always comply with such rules. Some of them do supply 'barbiturates' and 'opiates' that may look like the real thing, but aren't. This happens regularly with 'lifestyle' drugs such as Viagra or weight-loss medicines. Professional expertise is needed to establish whether the lethal drug that has been supplied is 'fake' or diluted (see the next section).

Other lethal drugs such as chloroquine (section 2.3) and tricyclic antidepressants (section 2.4) do not fall under international drug control conventions and are usually easier to obtain from local or Internet pharmacies. In the European Union, a prescription is required for these drugs. Not all (Internet) pharmacies comply with this requirement, however, as tourists in Southern or Eastern Europe have found out.

Can one trust drugs obtained over the Internet?

With the Internet comes the risk that one buys a fake drug. Therefore one needs help from volunteers of right-to-die organizations who, in the Netherlands, gather information about which Internet pharmacies are reliable and can sometimes give advice where to check the quality of the drugs that are delivered. They can also maintain a confidential list of reliable addresses and provide someone who is looking for a lethal drug with a single address. The Dutch right-to-die group called the Horizon (in Dutch: De Einder) checks whether the drug that has been delivered corresponds with earlier deliveries that had been proved to be reliable. When persons who have taken the drugs from a particular pharmacy have in fact died volunteers receive that information from relatives or friends of the deceased. In this way they can link the effectiveness of the drug to a particular supplier. By now some suppliers have proved to be reliable for several years because all who had ingested it have passed away.

Organizations from other countries who wish to further the cause of dignified dying might do similar work in their own countries. It would help clients if volunteers from different countries were to exchange this information with each other. Companies selling fake drugs could immediately be taken off the list of addresses of reliable suppliers. However, this exchange of information presupposes that volunteers trust each other. Since governments monitor e-mail-traffic, this kind of exchange may become more difficult.

How to take the drugs

Tablets should always be ground into a powder before being administered. Capsules can be opened up and the contents kept in a dry place till they are taken. Taking lethal drugs in powder form, with water, means that the substance dissolves before entering the stomach. A substance in tablet form will first have to dissolve in the stomach, so it takes longer to reach the blood, and it therefore takes longer before the heart stops or breathing ceases.

You can speed up the process of dissolving lethal drugs by sprinkling the ground-up tablets in yoghurt or custard, stirring it well, and immediately eating it with a spoon (if you wait, it will taste even more bitter). Do not speak while eating, as this helps to prevent choking. Drink one or two glasses of water, milk or fruit juice alongside. The drugs will then be absorbed more quickly in the blood.

Anti-emetics to prevent a failed attempt

Vomiting after ingesting the lethal substances is one of the main causes of failure, as the stomach may not be able to deal with the overdose of medicines. Two different anti-emetic medicines can be taken to prevent this. The first is metoclopramide (e.g. Maxolon, Primperan, Reglan), which is available on prescription only. Dose prior to self-chosen death: over a period of 24 hours, every 6 to 8 hours, take one 10 mg-tablet or two suppositories of 10 mg. This medicine is recommended as the most effective anti-emetic. The last tablet or suppository should be taken one hour prior to when you want to take the lethal powder.

The second choice is domperidon (e.g. Domperamol, Motilium), which is available from pharmacies without a prescription. Dose prior to self-chosen death: over a period of 24 hours, every 6 to 8 hours, take one 10 mg-tablet or two suppositories of 10 mg. The last dose should be taken about 1 hour before the set time of taking the drug.

Should the drugs be taken on an empty stomach?

One can continue to eat normally until twelve hours before taking lethal drugs. After that, we would advise you not to eat anything, so that the stomach is empty when the lethal drugs are taken. This encourages the quick assimilation of the drugs into the body. One can continue to drink water, tea or juice as normal. If you wish you may eat a light snack about 15-30 minutes prior to swallowing the lethal substances.

Interaction with other drugs that may hamper the lethal effect

Some medication can lessen the effect of a lethal drug. It is therefore recommended to stop taking medication, as far as possible, two weeks prior to the planned date. However, medication that has been prescribed for symptoms such as pain, shortness of breath or sickness should be continued. The same applies to medicines for cardiac arrhythmia, epilepsy or psychotic symptoms. Other medicines, including statins (that lower your cholesterol), anti-hypertensives, oral anti-diabetics and many others will no longer serve any purpose once the date of the self-chosen death has been fixed within the next week. In case of doubt, consult a doctor whom you can trust or a right-to-die group in your country (in the United Sates Compassion and Choices or Final Exit Network; in the UK Friends At The End or fate).

Alcohol: drink in moderation

Many authors have recommended the use of alcohol in hastening death. It is true that alcohol strengthens the effect of lethal drugs and of benzodiazepines. Unfortunately, the dose at which this effect occurs is not known and is likely to be quite high. The toxicologist estimates that this will only occur with five shot glasses of hard liquor, or up to 250 ml. Using this amount of alcohol entails serious risks. It may prevent the planned death from being carried out with the necessary carefulness. Moreover, alcohol can cause vomiting when one is not used to it – even after taking anti-emetics. The advice of the toxicologist is to use only the amount of alcohol that you are used to using and, preferably, to take the alcohol only after all lethal drugs have been taken.

When the person has fallen asleep: information for family and friends

Though death usually comes within 4 hours after the recommended dose has been taken it may sometimes take a long time for death to come. Especially with opiates or tricyclic antidepressants it may take longer as they slow down the passage through the intestines. Take turns to sit with the dying person, so as to prevent everyone from becoming exhausted. So long as the person is sleeping deeply, it is better not to call the doctor, even if it has been 24 hours. Occasional movements, moaning or irregular breathing are no indication that someone is waking up again.

If it does appear that the person might wake up, however, one can warn a doctor and ask whether they can give an injection of 10 mg of Valium (diazepam) every six hours in order to keep the person asleep. Some doctors are prepared to do this, because it is not a lethal act. Other doctors refuse to do this. The advance directive and the authorized person should make it clear to the doctor that the person should not be admitted to hospital. One need not worry if there is an interval of more than twelve hours, sometimes even more than 24 hours, before death occurs. People have woken with no lasting effects from very long periods of deep sleep (60 hours!). It can take several days to a week, however, before someone feels 'normal' again after waking up from such a long sleep.

In the USA death is reported through an emergency number. First an ambulance may rush in to resuscitate which should be prevented by the authorized person. Next the so called 'First Responders' will come in: the police and persons from the fire department who have the equipment to establish death. They will ask some standard questions (e.g. 'when did you last see this person alive'). If there are no suspicious signs, for instance medication boxes or inert gas equipment, the standard procedure is that a doctor who has not seen the patient himself will sign a declaration of death. Most often cardiac arrest is given as the cause.

In every country there are individuals who have experience with the phenomenon of a humane self-chosen death. Volunteers of right-to-die societies may be able to help you to get into contact with such a person in your neighborhood or can give information over the phone. He or she may be able to explain any unexpected events that may take place, and provide reassurance in the case that death takes more time than had been expected.

1.4 Summary: what one should know

1. Someone with a well-considered wish to die reaches a carefully weighed decision after consultation with loved-ones or friends; preferably also after consultation with a doctor or trained helper with experience in this area. Consultation will continue during the steps that follow. In the course of these discussions, it should become clear who you want to be present when your decision is carried out. This person(s) must be willing to do this and feel comfortable with being present at your death, especially in view of the possible legal consequences.
2. Method: a decision is made regarding the particular lethal medicine to be used, the sleeping pills to be taken with it, and an anti-emetic.
3. Procuring the ingredients: the lethal drug, at least one long-working benzodiazepine and the anti-emetic are procured. If the preferred drug is not available, one may choose another that is easier to obtain. Alternatively, one may turn to inert gas or to voluntary stopping eating and drinking.
4. Storing drugs: the drugs that have been procured are kept in a safe location that is not accessible to others. The drugs should be grinded in a coffee grinder prior to the planned date and sprinkled in yoghurt only just before ingesting them.
5. A refusal of life-saving treatment is written (advanced directive).
6. An authorized person is named. As soon as the date of the planned death is fixed, the authorized person is informed.
7. One should stop using other opiates and benzodiazepines for three weeks prior to the planned date. This is absolutely essential in the case of opiates, and it is recommended for sleeping pills. It is not necessary to stop chloroquine and antidepressants. Barbiturates are no longer prescribed on a regular basis.
8. Alcohol: Anyone wishing to use alcohol when taking the drugs should know how he or she reacts to alcohol. It is NOT necessary to take alcohol as part of a humane planned death.

1.4 Summary: what one should do

1. You should start to take anti-emetic tablets or suppositories about every 8 hours over the 25 hours before taking the legal drugs: one tablet or two suppositories. The last one should be taken one hour before taking the lethal drugs.
2. Eat very little or nothing during the last eight hours before taking the drugs. This is because medicines enter the body more quickly when the stomach is empty. You may drink as normal (e.g. water, milk, juice, coffee or tea).
3. On the day of the self-chosen death, get the following items ready for use at the intended moment:
   - the lethal pills and the sleeping pills;
   - a small bowl of thin yogurt (keep some extra yogurt in reserve); it is easier to take all these medicines when sprinkled in yogurt at the last minute (i.e. not beforehand);
   - a light snack of your choice, e.g. a biscuit or cracker with filling, might be eaten half an hour before you take the drugs.
   - two glasses of water or milk to wash away the taste and to facilitate the absorption from the stomach into your blood.
4. The authorized person is present at, or close to, the place of dying in order to prevent life-saving treatment being provided.
5. An incontinence pad can be used to prevent problems should the person become incontinent after falling asleep.
6. When all is said and done, sprinkle the ground-up lethal drugs and long-acting sleeping pills into the yogurt. Without delay, stir and swallow with a spoon. While eating, do not speak (to prevent choking). If some of the medicine mixture remains in the bowl, this can be taken with a little of the yogurt that has been kept in reserve.
7. Wash away the unpleasant taste with water, milk or fruit juice.
8. If a fast-acting sleeping pill has to be taken (in the case of chloroquine), the grinded pills should only now be sprinkled in the yoghurt, and eaten once all the other drugs have been taken. Otherwise you may become sleepy before you have finished eating the lethal drugs.

INTERMEZZO

Jan-Ru (78): a self-chosen death by pentobarbital[1]

Jan-Ru suffered from vascular dementia, which progressed slowly over a period of five years. His father had had the same disease and had ultimately died in a nursing home. Above all, Jan-Ru wanted to avoid that, but how? Helium? He couldn't do that by himself. Stopping eating and drinking? Not his thing. The least bad option seemed to be to search the Internet for a barbiturate.

Jan-Ru had always been a seaman, and after his retirement he sailed the Atlantic and the Pacific, together with his wife Nel. Nicknamed 'Piet', she was a former nurse, and had been his partner for 47 years. The couple had two sons and several grandchildren, with all of whom they had a good relationship. His wife describes him as 'a real alpha male, straightforward, determined to be in charge. Not a control freak, but a very good organizer.' In the film *A Way Out: Taking Control of Your Death*, she tells the story of her husband's self-chosen death.

'It all started with high blood pressure,' she recalls. 'He wouldn't see a doctor: "It came by itself, it'll leave by itself." Then he got a blood clot in one eye. Later his speech became slurred, and his energy gradually faded. I did everything I could to get him back on his feet, but gradually it became obvious that it was a lost battle. His joints stiffened, and he had to wear incontinence pads. But Jan-Ru never lost his sense of humor, and he was always immaculately dressed. He was able to wash and dress himself, but that was about all. As in many cases of vascular dementia, his speech was coherent and his memory remained largely intact.

'After the blood clot in his eye, he made it clear that he intended to manage his own death. "I don't want to end up in a nursing home." He'd hated seeing his father in an institution, and we were prepared to keep him at home.

'At first we decided on the helium method and ordered everything we would need. The boxes are still in the garage, so I can blow up plenty of balloons! Jan-Ru rejected helium, because it was too much trouble. He couldn't have done it himself anyway, since by that time he had lost all manual dexterity.

'We talked at length about stopping eating and drinking. But if you still enjoyed food and drink as much as he did, that wasn't an option.

'I was worried that he might become incompetent, and we discussed it with the children. He was particularly close to our younger son, as they often went sailing together. At one point he asked his father, "But why, Dad? Why?" Jan-Ru replied, "Paul, I can read one sentence, but after the second one I've forgotten the first. I can watch TV but I don't understand it anymore. So what's left for me?" Paul did all he could to talk him out of it: "Don't we have good times together?" he asked. "Yes, but for how long? Do I have to wait until I'm ready for a nursing home?" In the end, Paul was won over.

'We considered asking our family doctor for help. Jan-Ru had been her patient for years and she had followed his steady decline. Initially she said she wouldn't let him down, and promised to help him die with dignity. But when it came to the crunch, she said he hadn't reached that point yet. I was disappointed that she was led astray by his charm and his immaculate appearance. Who can decide for someone with vascular dementia how long he has?

'One day we heard about a right-to-die group. Since Jan-Ru could no longer cope with travel, we contacted a member of the group and he visited us at home. After we had gone through all the possibilities, we decided that the next step was to find a reliable address on the Internet. The counselor gave us the name of a supplier, and warned us that we would have to pay for our order in advance. There was no guarantee that it would arrive, since the customs people at either end might destroy the package without any notice.

'I had to do the ordering because Jan-Ru couldn't handle the computer any longer. I requested two 50-ml bottles of liquid pentobarbital, which to my relief arrived in three weeks. The counselor had strongly recommended that we also get 36 hours' worth of an anti-vomiting drug called metoclopramide. But at that time the supplier we used did not have it, and we could not get a prescription for it, so we had to just hope for the best.

'Jan-Ru said that he wanted us to enjoy this last summer together, and that on the first of September we'd have a glass of wine together.

'The children were informed, and our family doctor. She asked me whether I was going to pretend he died a natural death by a cardiac attack or cerebral accident.

'"No", I said, "That's not what I'm planning. After he passes away, I want to have one night with him alone. And then I'll inform the authorities."

'The counselor has explained how we can remain within the limits of the law. It's possible that the police will give us a hard time. But if it's all done secretly, the children and I will have to keep silent about it for years. And that's stressful too.

'On the day we had planned I phoned the family doctor and told her that it would take place that evening. She promised to come early the next morning to hand the medical file over to the coroner. The boys and their wives were with us and we had a lovely time out in the garden.

'We had agreed beforehand that they would stay outside while Jan-Ru and I went into the bedroom together. He looked at his watch and said: "It's eight o'clock." We hadn't finished all the food and drink but it was eight o'clock. "We're going to start now," he said, and then he went inside without taking leave of anyone.

'In the bedroom he sat down on the bed, then lay down half propped up, the way we'd been advised.

'We looked at each other. Everything had been said. I poured the mixture into two glasses. He'd asked for two bottles. First he dipped one finger into the liquid and tasted it. Apparently it was all right. Without a word, he then finished the two glasses, one after the other.

'He said a few words that were inaudible. For an instant he seemed short of breath. I was afraid he might vomit, so I put the bed straight up. But that passed and after five to seven minutes at most, he was gone.

'What can I say? Everything had already been said. It felt so unreal. So unreal.

'Then I told the children. They all went into the bedroom to say goodbye, and then I called our doctor. I said all had gone well, which was true. I know it sounds awful, but it did go well. Paul stayed with me and slept in the living room. I simply lay down beside Jan-Ru. To be honest, it didn't bother me to fall asleep next to him. In fact, it was a good feeling.

'At eight in the morning our family doctor arrived as promised. She telephoned the coroner, who informed the police. The doctor had asked if they could come in civilian dress and in unmarked cars. But according to protocol, it had to be uniforms and police cars.

'It was gorgeous weather. I opened the front door and said, "Good morning gentlemen."

'"Doesn't sound like a very good morning," one of them muttered.

'"Look up at the sun, sir. This is a good morning."

'They shuffled into the house. I introduced them to our family doctor and then it was a matter of waiting for the next lot. Three different groups of people came and went. Each time someone asked what had happened, I told them that my husband had swallowed a lethal mixture and died. I could tell they thought this was something that shouldn't have been done. But it had been done. They also asked when he had died, and I told them "last night."

'Then yet another group of people arrived, and it was becoming distressing. One man, who turned out to be the coroner, glanced through the medical file, reading about Jan-Ru's illness and how we had coped with it for years. There was also a police detective, a woman, who asked me how I'd got hold of the medication. I said: via the Internet. She asked me for the Internet address, but I refused to give it. She didn't press the point, because our doctor (who had stayed in the room) looked up and said, "Just read that book over there, *A Way to Die*. Then everything will be clear and you won't have to ask all these questions." Our doctor definitely made it easier for us, ironing out difficulties with the authorities.

'How do I look back on all this? I wouldn't like to have missed those last few years, you know. When he had his first stroke, quite a while before, it would really have been too early. I'm happy, truly happy, that things went the way they did. And that we had those last good years together. This is a way of dying that you would wish for everyone: in your own time, in your own way, in your own bed.'

Chapter 2

Lethal drugs

2.1
Barbiturates

Barbiturates are substances that fall under international drug control conventions. There are strict controls on the sale, transport and possession of these drugs; see section 1.3 for more information on how to get hold of them. They are used in anesthesiology, palliative care and veterinary medicine. Physicians who prescribe barbiturates for other purposes risk losing their license.

This section only provides detailed information on pentobarbital and phenobarbital. There are many other barbiturates, such as secobarbital, thiopental, amobarbital and butobarbital, but we do not cover them because they are too difficult to obtain.

Ever since the 1970s, when the Dutch anesthesiologist Pieter Admiraal shared his expert knowledge of barbiturates with Derek Humphry, barbiturates have been used to ensure a peaceful death. More recently, Philip Nitschke made a bold but unsuccessful attempt to circumvent the law by synthesizing a kitchen version of a barbiturate, using freely available chemical substances.

To obtain a lethal dose of a barbiturate, the very old and the very ill for whom life is no longer tolerable must rely on the help of their children or close friends (the next generation of elderly people will be less dependent). It is really difficult to find one's way through the jungle of Internet sites that sell barbiturates. A coordinated effort is therefore needed on the part of volunteers of members of right-to-die organizations to search for reliable Internet pharmacies.

It is of no use to publish these addresses. The authorities might block access to them or might sue these companies, which may result in their addresses changing or disappearing altogether.

Another problem is how to ascertain whether barbiturates delivered by Internet sites have been cut or polluted. It is advisable to use addresses that have been shown in the past to deliver pentobarbital (as a powder or liquid) with a purity level of at least 95%. The first ten or more samples delivered should be tested in a specialized laboratory.

In Australia, Philip Nitschke has claimed that he is able to test pentobarbital powder from China in a small chemical laboratory in a van. According to the pharmacist and the toxicologist I consulted, who have quite some experience with these tests they are adequate for a qualitative analysis to ensure that the powder is not largely fake. No detailed information has been given by Nitschke on these tests. In toxicology only gas chromatography is considered adequate for a quantitative analysis of the sample. To date, this very expensive method has not been mentioned in Exit International newsletters. The specialists were unable to confirm that Nitschke's tests accurately report levels of 90% purity or more.

It would be a great step forward if specialists from English-speaking countries and European countries where the law is more tolerant were to work together to determine the purity of pentobarbital samples. An open exchange among professionals about the tests used would be necessary.

Couple suicides with barbiturates are becoming a hot issue in the Netherlands. When both reach their eighties and one of them gets a serious disease the other sometimes prefers to die together instead of living on alone with all sorts of disabilities. Often the children are informed about the wish of their parents to leave the world hand in hand. This romanticized picture is worrying because subtle psychological pressure from the one on the other is not rare in my experience.

Pentobarbital

Please note: in the boxes below, technical terms are used that are all explained in chapter 1.

Liquid pentobarbital is called Nembutal. Nembutal from Mexico usually comes as a liquid in a bottle containing about 6 grams. Pentobarbital from China usually comes as a powder in different quantities. In the near future online companies in India or other non western countries may deliver pentobarbital as well. It will take a coordinated effort by volunteers of several right-to-die groups to develop reliable contacts in these countries.

Table 2.1 Nembutal and Pentobarbital

Brand names	Nembutal (or liquid pentobarbital) for oral use or injection is usually sold in bottles of 100 ml that contain 6 g. Pentobarbital will come in tablet or powder form.
Availability	Can be ordered from Internet pharmacies known to experts in right-to-die organizations. Used to be for sale without a prescription from vets in Mexican towns.
Sleeping pills	Not necessary. When taken in the recommended dosage, all barbiturates produce a deep and long-lasting sleep.
Cause of death	Cardiac arrest and depression of respiration.
Lethal dose*	6 g is lethal. In Switzerland, Belgium and Holland, 15 g is prescribed by doctors only to speed up time to death.
Time to death**	With 6 g it varies from 5 minutes to 48 hours in rare cases. Don't panic and don't call a doctor.
How to take it	Nembutal (liquid): pour it into a glass and drink it, followed by a glass of water, milk or juice. Vets inject it into a muscle, Dutch doctors use the intravenous route. Pentobarbital (powder): sprinkle powder in thin yoghurt. Eat it followed by water, milk or juice.
Storage life	Liquid pentobarbital: 3 years if properly sealed and stored in a dark place at room temperature. (Don't store it where others may find it.) Powder can be stored in the same way for 5 years or more.
Detoxification	No detox period is required.
Anti-emetics	Advised at about 13, 7 and 1 hours before ingesting.

* In 2012 the Royal Dutch Association of Physicians (KNMG) and the Royal Dutch Association of Pharmacists jointly published a new Guideline for physician-assisted dying by doctors (see Appendix 5).
** Time to death: data from the Oregon Department of Human Services (March 2006) indicate that 9-12 g of pentobarbital taken orally leads to death in about 25 minutes (based on 200 cases, ranging from 4 minutes to 48 hours). Loss of consciousness occurs on average after 5 minutes (range 1-38 minutes). These data can give some idea of what to expect.[1]

Table 2.1 continued

Combination with other medicines	Stop laxatives 5 days before. It is said that an overdose of a beta-blocker may speed up death. We discourage beta-blockers as they might also delay death.
Reported cases	Countless. Only one reported failure in Oregon.
Disadvantages	The powder or liquid may be diluted or mixed with unknown substances. It is advisable to use a supplier whose products have proved to be reliable. Barbiturates taste bitter and soapy; drinking something sweet (syrup) after ingesting them helps to mask the bad taste.
Customs	All barbiturates are controlled substances. If discovered by customs, they will be destroyed. To date, no addressee has been prosecuted for ordering or possessing Nembutal.

Phenobarbital

There are only a few reliable reports on the use of phenobarbital to end one's life. Therefore it will be mentioned in Chapter 4.3.3 among the other medicines that might cause a humane death. Its lethal potential has not been proven in about 20 reported cases.

Secobarbital

In the USA secobarbital is still prescribed for patients with persistent severe complaints of sleeplessness. One needs a good story to get them prescribed in small quantities.

Except for availability all information for secobarbital is the same as for pentobarbital.

2.2
Opiates

Opiates are substances that fall under international drug control conventions. There are strict controls on the sale, transport and possession of these drugs; see section 1.3 for more information on how one might get hold of them.

All opiates cause death by depression of respiration or 'apnea'. Every opiate causes habituation to all other opiates, so one should check carefully whether one of the medicines prescribed in the case of pain contains an opiate. If so, using opiates for a self-chosen death will be very risky, unless a detox and three-week 'clean' period are observed.

Morphine
There are only a few reliable reports on the use of morphine to end one's life. Therefore it will be mentioned in Chapter 4.3.3 among the other medicines that might cause a humane death. Its lethal potential has not been proven in about 20 reported cases.

Fentanyl (table 2.2)
Fentanyl in transdermal patches might be lethal. Fentanyl in sublingual drops or buccal spray might be lethal in theory but it is not lethal in practice, as it would be hard to collect enough of it.

Oxycodone (table 2.3)
Oxycodone is an opiate that is almost twice as strong as morphine. It is available as coated tablets with a hard outer shell, or as capsules that can be opened to release the contents. Oxycodone is also available in liquid form for subcutaneous injection.

Dextropropoxyphene or D-propoxyphene (table 2.4)
This opiate was taken off the market in most European countries in 2012. For someone who obtained it, the information in the table may be useful.

Table 2.2 Fentanyl

Brand names	Durogesic, Fentanyl.
Availability	On prescription.
Prescribed	For intractable pain, especially in the case of terminal cancer. With effort and patience, sufficient plasters can be hoarded. Some Internet pharmacies sell them.
Habituation and detoxification	Very rapid. Other opiates taken orally or by injection also result in habituation. Detox and a clean period of 3 weeks are required. In the case of pain, this may be impossible.
Lethal dose *	Uncertain, due to inaccurate reports. To be sure, experts advise 500 mcg, e.g. 20 patches of 25 mcg .
Sleeping pills	Necessary: 500 mg of a long-acting benzodiazepine after fixing the patches.
How to apply *	Fix the patches on skin that is free of hair and apply warmth.
Time until death	12 to 60 hours.
Reported cases	The US Food and Drug Administration reports over 100 cases per year. The majority are accidental deaths where the patches have been heated by a hot compress, an electric blanket or other means.
Disadvantages	Procuring them remains a problem. In cases of habituation to other opiates, it may take several days before apnea occurs.

* Fentanyl is released from the plaster into the blood more quickly if warmth is applied by means of a hot compress, an electric blanket or a hot water bottle near the plasters. 300 micrograms may then be sufficient.[2] The US Food and Drug Administration reports more than one hundred fentanyl deaths per year. Most of them are accidental deaths due to warmth near the plaster.

Methadone and Heroin
There are only a few reliable reports on the use of Methadone or Heroin to end one's life. Therefore they will be discussed in Chapter 4.3.3 among the other medicines that might cause a humane death. Their potential lethal action has not been proven in about 20 reported cases.

Table 2.3 Oxycodone

Brand or trade names	Oxynorm, Oxycodon. Note: OxyContin contains 'slow-release' oxycodone that enters the blood too slowly and may not cause death. We advise against the use of OxyContin.*
Availability	Via a doctor's prescription or an Internet pharmacy.
Prescribed	By doctors for severe and persistent pain.
Habituation and detoxification	Rapidly and for all other opiates as well. Detox and 3 weeks 'clean' are essential.
Lethal dose	1 g in tablets, or as a drink.
Sleeping pills	Necessary: 500 mg of a long-acting benzodiazepine (flurazepam or diazepam).
How to take	The tablets are ground and then swallowed together with the sleeping pills. If in liquid form, it is better to take the sleeping pills first.
Time to death	1-48 hours.
Reported cases	Over 40 people have died in the Netherlands after taking 1200 mg of oxycodon with 120 mg of Flunitrazepam (Rohypnol). An attempt with 1200 mg of Oxycontin, i.e. slow-release tablets, failed.*
Disadvantages	A controlled substance, transport is not permitted. Habituation to all other opiates.

* Report: a 72-year-old woman took 1200 mg of Oxycontin but continued to live. The family notified the doctor, who had her hospitalized. Several days later she regained consciousness, having sustained no permanent damage. Explanation: she had taken OxyContin that contains 'slow-release' oxycodone, which enters the blood too slowly to stop respiration. I advise against Oxycontin.

Table 2.4 Dextropropoxyphene or D-propoxyphene

Dextro propxyphene	Depronal, Deprancol, Darvon, Propoxyphene.
Availability	From some Internet pharmacies in developing countries.
Prescribed	For serious pain, especially due to rheumatoid arthritis and cancer.
Habituation and detoxification	Like other opiates, very rapid.
Lethal dose	4.5 grams, i.e. 30 capsules of 150 mg
Sleeping pills	Necessary: 500 mg of a long-acting benzodiazepine.
Use	In capsule form, the drug is coated in granules for slow release. The capsules should be opened and the contents ground in a mortar. Tablets should be ground.
Time until death	8-48 hours.
Reported cases	90 successful and 3 failed.
Disadvantages	Controlled substance. Transport is not permitted Habituation to all other opiates.
Reported failures	– A patient with disseminated cancer attempted to commit suicide using dextropropoxyphene, but did not notice she used a painkiller in the opiate group. As a result, she forgot the detox, so the dextropropoxyphene was not lethal. – A patient purchased capsules in a developing country that turned out to be fake. – A 90-year-old woman took only twenty capsules, instead of the recommended thirty. She did not die until 43 hours later, probably due in part to dehydration.

A note on anti-depressants with a tricyclic chemical structure
According to the toxicologist Pennings, in appropriate doses these drugs are very likely to cause death by cardiac arrest. They will be discussed in chapter 4.3.3 because less than 20 cases have been reported in which they were used for a self-chosen and dignified death.

2.3
Chloroquine

An overdose of chloroquine compounds will result in cardiac arrest. There are three compounds. Two of these are anti-malaria drugs: chloroquine sulfate and chloroquine phosphate. The third compound is hydroxychloroquine, which will be discussed in chapter 4 because less than 20 cases have been reported in which it was used for a self-chosen death. It is prescribed for rheumatoid arthritis.

Availability
Chloroquine compounds do not fall under international drug control conventions and are therefore usually easier to obtain than opiates and barbiturates. Doctors in Western countries may be reluctant to prescribe them as most mosquito's have become resistant to chloroquine. In countries where malaria is still endemic, they can be obtained from pharmacies without a prescription. Internet pharmacies will often sell them as these drugs do not fall under international regulations.

Sleeping pills
Someone planning to end his or her life using chloroquine will need to take two types of sleeping pills: a long-acting benzodiazepine (at least 500 mg) at the same time as the chloroquine, and immediately after that, a fast-acting benzodiazepine (100 mg). This is because an overdose of chloroquine may quickly lead to painful muscle spasms or an epileptic seizure. At that point, the concentration of the long-acting narcotic in the blood is not yet sufficient to suppress these side effects. The fast-acting sleeping pills will suppress such side effects, but after a few hours they lose their effect. By that time, the concentration of long-acting sleeping pills will be sufficient.

Observations and information from the literature
Docker (2013) has argued that diazepam reduces the lethal effect of chloroquine. This hypothesis has been refuted by the observations of Dr. Uwe Christian Arnold from Germany, who attended over 50 cases of self-chosen death by chloroquine phosphate (Resochin) and 1000

mg of diazepam (Valium). All cases died within two to four hours. One should note that in Germany, the kind of assistance that Dr. Arnold provided is not against the law.

Cases of chloroquine overdose that receive treatment in hospital are given diazepam intravenously to suppress muscle spasms and epileptic seizures.[3] According to toxicologist Ed Pennings it has not been convincingly demonstrated in the toxicological literature that diazepam blocks the toxic effect of chloroquine on the heart. Studies on animals[4] have reported results that conflict with each other. Docker admits in his 2013 edition of *Five Last Acts* that a massive overdose of chloroquine (see doses mentioned in the tables below) will cause death irrespective of the sedative that is taken.[5]

Hydroxychloroquine will be mentioned in chapter 4.3.3 because less than 20 cases have been reported in which it was used for a self-chosen death.

How to take chloroquine and sleeping pills

Before the planned date, grind the chloroquine tablets in an electric coffee grinder or a mortar. Store the powder in a well-sealed bottle. Follow the same procedure with the long-acting and fast-acting pills, and store them separately.

On the appointed day, sprinkle the chloroquine powder and the long-acting sleeping pills together into a cup of thin yoghurt. Do the same with the fast-acting sleeping pills in a separate second cup. Stir well and finish the first mixture without delay. The longer you wait, the more bitter it will become. Then finish the second cup with the fast-acting sleeping pills. After emptying the cups, you can drink one or two glasses of water, milk or juice. You may then eat something sweet to mask the bitterness, if you'd like to.

Since chloroquine is extremely bitter, an anti-emetic is essential.

Table 2.5 Chloroquine phosphate

Brand names	Resochin
Availability	Internet pharmacies. In many countries outside the EU and the USA, drugs can be obtained without a prescription. Caution: drugs bought over the Internet may be fake.
Prescribed	For malaria, in regions where the parasite has not yet become resistant to chloroquine.
Habituation	None. It is recommended that one is 'clean' from sleeping pills for two weeks.
Lethal dose	17.5 g, equal to 70 250-mg tablets.
Sleeping pills	It is essential to take 100 mg of fast-acting benzodiazepines, as the drug is absorbed quickly and may cause spasms within an hour. To maintain a deep sleep, 500 mg of long-acting BD is necessary.
Procedure	Crush them in an electric coffee grinder and sprinkle the powder, together with the long-acting diazepam, over yoghurt. The fast-acting sleeping pill may make you drowsy very quickly, and should be taken last.
Time to death	From 4 to 24 hours.
Reported cases	Over 50 reports from a physician who was present. No instances of failure.
Side-effects	Muscle spasms and epileptic seizures occurred in 2 people who did not take diazepam (which will suppress this).

Table 2.6 Chloroquine sulphate

Brand name	A-CQ chloroquine
Lethal dose	11 g; 110 tablets of 100 mg.
Reported cases	20 reports, no instances of failure.

All other information is the same as in the table for chloroquine phosphate.

INTERMEZZO

Annelies (76): a self-chosen death with helium[1]

Death by helium? 'Undignified', say many people in the Netherlands. 'Humane', says Catharina Vasterling, a board member of the Dutch Foundation for Dignified Dying. She recounts how she witnessed the helium exit of her friend Annelies. She was joined in witnessing the death by Els, a cousin of Annelies. Els relates her experience in the film 'A Way to Die. Eyewitnesses on Methods for a Self-Chosen and Humane Death.'[2]

"'I've known Annelies for over 30 thirty years. We met at university, where she taught nursing. She never married and had no children, but her circle of friends was wide. 'Annelies was always in charge. She was an intelligent and practical woman, interested in all the latest trends: computer games, iPads, and gadgets. She was very conscious that her life would soon draw to a close and sometimes made jokes about it.

'Annelies suffered from COPD (chronic obstructive pulmonary disease), which caused tightness in her chest. She was given a new hip, but walking was still difficult, and she was frequently in pain because of her rheumatism.

'Annelies was constantly shifting her pain barrier. A heart operation had been performed, but the improvement lasted only a year. In 2012 she was rushed to the hospital with pneumonia on three separate occasions. Annelies had appointed me as her proxy decision-maker, in case she was unable to decide for herself whether or not to have the pneumonia treated. She wasn't afraid to die, but she had a fear of suffocating. During her last few months she was extremely short of breath, and after her release from the hospital she was on oxygen and morphine. Her severe exhaustion was visible. The year before, she could still pull algae out of the pond in her garden, but that activity was now too tiring for her.

'It was clear to me that she was suffering, and that there was no likelihood of improvement. Her bed had been moved to her living-room, since she could no longer manage the stairs. She wanted a view of her garden, which she dearly loved.

'Annelies knew that her doctor would never agree to provide physician-assisted dying. Years before, I had persuaded her to change doctors, which she did. But later she returned to her former doctor because she had such a good relationship with him. He was willing to relieve her pain, but giving her barbiturates was not in keeping with his personal moral philosophy.

'After the third bout of pneumonia within a year, Annelies wanted nothing more than to go home and simply stop eating and drinking. But she gave up that idea, as she was reluctant to burden her friends with the intensive nursing that would be necessary. Those close to her would have to wait for one or two weeks until the end came, and she could not abide the prospect of lying there in a kind of coma.

'Unable to make a decision, she saw her life dragging on, while her condition worsened and waiting for the end became unbearable.

'Annelies decided to collect chloroquine and valium. This is the lethal cocktail that has been described in chapter 2 of this book. I ordered them for her online. When she saw the number of pills, she was afraid she'd never manage it and would have to give up halfway through.

'Shortly afterwards Annelies heard about the End of Life Clinic, an initiative of Right to Die NL. Anyone who has been turned down by his or her own physician can register there for a second opinion and possibly receive physician-assisted dying from one of the doctors who work in this new organization. She placed all her hope in these people, but the waiting list was so long that it would be over six months before her request could be evaluated. She began to lose hope for a way out that would suit her.

'At that point I told her that there was a film about the helium method for a humane self-chosen death. I'm a board member of the Dutch Foundation for Dignified Dying, and we circulate a film about exiting via helium. Many people regard it as undignified to die with a hood over your head, which is what happens with the helium method. But when I saw the film, I was surprised by how simple the method was and how gently death comes. I thought to myself: Hey, this is something I wouldn't mind putting my name to.

'I did ask myself whether the helium method isn't a bit too easy for young people who are suicidal for some reason. But although you can indeed do it by yourself, there's a lot of preparation involved; it certainly requires more preparation than hanging. And it is more humane than throwing yourself in front of a train.

'Annelies ordered the DVD and watched it several times. Gradually she got used to the idea of putting a bag over her head. The required materials are not that hard to find. The two helium tanks can be ordered from the Party Shop or via the Internet, and the large plastic roasting bag is also available on the Internet. You need elastic and tape as well. The hardest part was finding the plastic tubing through which the gas flows from the tanks into the bag.

'First you have to practice the whole routine. This is not something you do on impulse. It's important to keep practicing until you know what you're doing. Annelies went through the routine some twenty times, and often I was there to lend a bit of moral support.

'Annelies was determined that nothing would go wrong. She practiced it over and over, but without the gas. During the last few months the tanks with the bright pink balloons were permanently ensconced in her living-room. She worked out the best way to grasp the bag and to press the air out of the bag, which was perched on her head like a shower cap. She then pulled it over her head in a single movement. In another session without the bag over her face she opened the tap on the tank until the bag was bulging with helium. When the time came, she would pull the bag full of helium down to cover her face. In the beginning it took a lot of fiddling, but ultimately it became a smooth automatic motion.

'On April 23 Annelies announced that she wanted to set the date. She decided on the ninth of May, the day I'd be coming back from vacation. Her niece would also be present. In the meantime, her death was constantly in my thoughts. It was an emotional rollercoaster and there were swings in her determination to go through with it. On a good day, life didn't seem that bad to her.

'I suggested that it might be helpful to stop the morphine, to get some sense of how she would feel without it, and to find out if she was still determined to go through with her plan. What worried her most was the thought that at the last minute she might change her mind.

She also began to suffer memory lapses. But when we asked her: '"Is this still what you want?",' her answer was '"Yes."'

'The first of May was Annelies' birthday. She celebrated it with her niece and other good friends of hers, and she clearly enjoyed herself.

'The following week her doctor came by. He was aware of what was going to happen, and had previously given her a calming medication. That is not against the law. During those final days Annelies was nervous. Her niece arrived the evening before. They discussed the details. Things were settled.

'On the day Annelies had chosen for her death, I went to her house, where her cousin and three women female friends had gathered. We had coffee and a glass of wine. Then Annelies announced that it was time, and sent the women home. It was not an emotional farewell. Her niece asked, '"Are you really sure?"' Annelies said, '"Everything's been said."' She even joked about the situation. '"Just look at me – everyone has to die sometime and here I am, making a big production out of it."'

'Her niece and I sat on either side of the bed. Annelies settled herself between two piles of pillows, so that she wouldn't fall over when she lost consciousness. And then she said goodbye. For the last time, she performed what she called her '"trick"', the one she had practiced so often. She breathed out and with both hands pulled the bag – now full of helium – down over her head.

'Three breaths, and she was gone. We saw no sign of distress. We had prepared ourselves for muscle spasms, but there were only three slight twitches. And there was nothing macabre about it.

'After half an hour we called the doctor, who was waiting to hear from us. We had calculated the risk and asked ourselves what we would do if things went wrong. We didn't have a Plan B. But I think we would have helped her. How, I'm not sure. Fortunately it wasn't necessary. According to the Dutch Supreme Court decision in 1995, you may lend moral support by not making the person die alone. But you may not say, '"Now do this or that,"', and by no means are you allowed to provide active assistance leading to death.

'We knew that the police always visit the scene of a non-natural death. Her doctor contacted the coroner, who established the cause of death. He arrived with four policemen in his wake. Two police officers took up their position at the front and the back doors, and two others stood guard over us. The doctor told the coroner what had happened, and I showed him the farewell letter that Annelies had written, in which she made it clear that, despite the morphine, her life was no longer worth living. She also made it clear that she had brought about her own death, consciously and after due consideration. The coroner's impression was that nothing illegal had taken place, but he asked us to wait until the detectives arrived.

'We had expected it would all be over in a half-hour or so and we could go on to lay her out. But in fact we had to wait several hours. First there were two detectives, followed by a forensic expert and then a public prosecutor. Imposing these long hours of waiting seemed disrespectful to us, and we were disturbed by the way our "guests" wandered all over the house while Annelies was still lying there in bed. Their presence also meant that we were unable to lay her out and wash her.

'We were questioned separately about what had happened. They wanted to know how she managed to take her own life with such rheumatic hands, so I told them she'd practiced so often that it was no longer a problem. They didn't ask whether I'd helped by buying some ingredients and I didn't tell them.

'At first the assistant public prosecutor refused to release the body. She said we were suspected of failing to rescue someone who was fighting for her life, and that there is an article in Dutch law which makes this a punishable offense. Were we supposed to pull the bag off her head? The situation was becoming tense. But then I showed them the prohibition of treatment that Annelies had drawn up, which made it clear that she did not want to be "'rescued'". There were phone conversations with their superior and we were cleared of any crime.

'Her body was then handed over for cremation. However, rigor mortis had already set in, making it impossible for us to care for her ourselves. That's what did make me angry. But on the other hand, I had taken into account that we might well end up in jail, since there is no case law yet by the Supreme Court clarifying whether loved ones are allowed to assist in exit preparations.

'The following Saturday Annelies' body was placed on a bier in her own home, where friends came to take leave of her. There was a small group of people present at the crematorium, and we accompanied her to the oven. She had made it clear in the final months before her death that she did not want a ceremony. But she did draw up a list of contacts, to whom she wished to leave certain belongings. That was Annelies: in charge. And things went as she wished.'"

Chapter 3

Inert gases: Helium or Nitrogen

3.1
Introduction

Helium gas

Since 1999 helium gas in balloon kits has become the preferred method for a self-chosen and humane death in some right-to-die organizations. In April 2015 the largest manufacturer of disposable helium cylinders, Worthington in the USA, has announced that from April 2015 their helium cylinders ('Balloon Time kits') will guarantee only 80% helium, with up to 20% air. This has been motivated by considerations of cost-reduction and the expectation of a shortage of helium gas.

While an 80/20 helium with air mixture is suitable for floating party balloons, its use to provide a peaceful death is almost certainly lost. At the NuTech meeting of June 2015, most experts agreed that the presence of about 4.2% oxygen in the new generation of tanks may perhaps be enough to cause unconsciousness but death may follow after many hours or, perhaps, not at all. The dilution must be indicated on the tanks. Anyone who purchases helium for a self-chosen death should carefully check this. The old tanks with almost pure helium may well be around for some time. (See *Balloon Occasions,* manufactured by BOC, a gas company).

Nitrogen gas

Richard Cone MD PhD has suggested improvements of the information on nitrogen gas that hat I have accepted with gratitude.

The dilution of helium tanks with air has prompted right-to-die organizations to search for alternative routes to a dignified self-chosen death. Philip Nitschke, director of Exit International, has proposed that helium should be replaced by other inert gases such as nitrogen or perhaps argon gas.[1] In Australia Nitschkes company Max Dog Brewing distributes tanks with almost pure nitrogen for improving the taste of beer. Nitschke has reported that over a hundred persons in Australia have died by inhaling nitrogen from a plastic bag over the head.

Though it may be likely that inhaling pure nitrogen gas causes death it has not yet been established that death comes as quickly as with helium gas. With helium there is enough evidence (on video and from professional observers who have witnessed helium deaths) about the time it takes to lose consciousness first and then die.[2] The substitution of helium gas by nitrogen for a quick and painless death is in need of observational reports, preferably recordings on video that can be analyzed by professionals.

In the USA and Europe shipping of nitrogen tanks is at present being researched. Pure nitrogen is used in welding and therefore can be acquired in welding shops. Any layperson can buy a full 20 ft3 tank in a welding store or online from a company. Apart from a full tank one needs an inert gas flow regulator as the pressure in nitrogen tanks is much higher than in helium tanks. This can be purchased separately at a welding or hardware shop.

Some mistakenly claim that nitrogen is 'totally undetectable'. Ogden gives references to the forensic literature on inert gases. Both nitrogen and helium gas can be detected with specialist techniques in postmortem blood samples.[3] The authorities in the UK have become increasingly aware of the use of gases for suicide (see appendix 4).

In the remainder of this chapter I will discuss helium and nitrogen. Pure helium gas is still around over the internet or at party balloon shops while the acquisition of nitrogen requires some effort.

If pure helium is not available the only extra effort in case of nitrogen is that a flow regulator is needed. Many elderly or ill persons will need help to fix this properly. They will also need some practice how to turn the tap on the tank (see section 3.4).

3.2
Legal issues

In the introductory chapter it has been discussed that some very ill or very old persons are physically not able to perform the preparations for the inert gas method. In the Netherlands these preparations are not considered a punishable offense provided there is clear evidence that the deceased had a persistent and well considered wish to die. Moreover, the acts that together cause death should be done by the person him or herself: turning the tap on the tank and pulling the bag over one's face. This evidence can be recorded on video.

Depending on the law in other countries or US States these recordings might give some protection in case of a police investigation. Some jurisdictions are so repressive that even being present without providing any help will be considered a punishable offence, specifically aiding and abetting a suicide. Relatives who are determined not to let a close relative die alone will be forced under those circumstances to be present in secret and immediately remove all ingredients after death.

3.3
Emotional preparation

At some point, people who are elderly and seriously ill may say, 'I've had enough'. If you probe a little, however, there may be many things that they want to do before letting go. They have not reached the point where it would be all right if the end were to come tomorrow.

Accepting death is contrary to human nature. We frequently move all kinds of milestones that we'd set in the past ('I'd rather die than…'). There is often inner conflict between our desire to stay alive for just a bit longer, despite of all the suffering, and our wish to die.

Even though you feel torn, however, you may still decide to make concrete arrangements for a humane death, to be prepared when the time comes. The inert gas method is rapid, painless, and safe. But before embarking on such preparations, a choice must be made. What do you dread more: living the final part of your life in dependency, or dying a bit sooner, while you are still able to do things for yourself?

Whether you opt for the medicine method or inert gas, <u>you cannot wait until you're too weak to get out of bed</u>. Voluntary stopping eating and drinking can be postponed, provided your brain is still active and good care can be arranged. But for the inert gas and medication methods outlined in this book, preparations must be made. If you keep putting off making the necessary arrangements, there is the risk of losing the ability to take control of your own death.

After the preparations have been made, they needn't be put into practice until the balance shifts to a clear sense of 'enough is enough.' As long as you keep saying contradictory things or changing your plans from one day to the next, you are not yet ready to implement those plans. Choosing the right moment will always be a difficult decision, from the standpoint of the person concerned and all those who are close to them.

The KISS Principle: Keep It Simple, Stupid

The physician Richard MacDonald has stressed the 'KISS principle' for the inert gas method: Keep It Simple, Stupid. He argued that the ill or elderly should not meddle with technical issues that would only deter them from making use of this effective, painless and perfectly legal method. Dying by inhaling helium or nitrogen should be as simple as possible. No method is 100% foolproof, and we have to live with that very small chance of waking up. If that happens, you can seek expert advice on what went wrong, and correct your mistakes. In the few cases of failure, no permanent brain damage has been reported, neither to me nor in the literature.

Textbox 3.1 Why an inert gas instead of medication?

Many people prefer the medication method. However, an inert gas is an option for those who:
- cannot obtain lethal medication;
- do not want to order an illegal substance;
- want a faster way to die than using medicine;
- cannot break the habit of taking painkillers and sleeping pills;
- do not have the motivation and willpower to hasten their death by stopping eating and drinking

---

What is special about helium gas?

Helium tanks can be obtained from party shops, toy stores and the Internet. Helium comes in small, non-refillable tanks for inflating party balloons. Helium is colorless and odorless, and is neither explosive nor flammable. When someone breathes pure helium from a helium tank connected to a plastic bag or a hood over one's head, this almost immediately results in death through oxygen deprivation. The patient experiences no sensation of suffocation. How is that possible? Normally, it is the carbon dioxide exhaled from the lungs into the bag that causes a sense of suffocation. The flow of helium from the tank drives carbon dioxide that is exhaled out of the bag.

Nitrogen gas tanks can be obtained from a welding shop or through gas companies on the Internet (see appendix 6). There is no precise information yet whether nitrogen gas causes death as quickly as helium. On theoretical grounds no significant differences with helium are to be expected.

Neither helium nor nitrogen is detectable during a standard post-mortem examination. However, methods for the detection of these gases in post-mortem blood and tissues have been developed.

Neither helium nor nitrogen is dangerous to anyone sitting in the room. In any average size room an inert gas quickly disperses and presents no danger. It is only when breathing pure helium or nitrogen that a person is at risk.

Observations

The Canadian criminologist Russel Ogden has published his observations of several deaths due to helium inhalation. After the deaths, he reported them to the coroner and the police.[4] His detailed accounts were confirmed to me in person by Dr. Uwe Arnold from Germany and an American doctor, each of whom had attended over a hundred helium deaths. The consistency of these reports from different sources convinced me that helium is effective, painless and safe for all those present.

Ogden collected reports on 119 persons who had died by means of helium. In 62 cases, the time to unconsciousness was reported: the average was 35 seconds (range: 10 to 140 seconds). The elapsed time to death was reported in 108 cases. The average was 13 minutes (range 2 to 40 minutes).[5]

These findings match those of Horikx in cases where a physician gives two injections: first with thiopental (2 g), which causes deep sleep; second with propofol (150 mg), a curare-like drug, that paralyzes the respiratory muscles. In the 134 cases that were reported, death resulted within 15 minutes.[6] The research data from Horikx and Ogden demonstrate that by far the best way doctors have got to cause death without pain, quickly and effectively, is matched by the speed with which an informed layperson can do the same by inhaling an inert gas.

Frequency in the Netherlands and UK

In the Netherlands, information about the helium method has become available recently.[7] To date, there has been no survey study of the helium method for a self-chosen humane death. In 2013, in a province of the Netherlands with about 2 million inhabitants, 17 helium deaths were reported by coroners. Some of these were lonely deaths by helium; relatives and friends may have been involved in other cases. In the UK and Ireland (53 million people) 53 helium deaths were reported in 2011. In San Diego County (population 3 million) the number of reported helium deaths rose in 2011 and 2012 to 19 cases. In Australia (population 23 million) only 79 self-chosen death by helium have been registered over a five year period (2005-2009).[8] For further information see the data in appendix 4.

Films

3.4
Preparations

Textbox 3.2 Two films on the inert gas method

If the description of the procedure below is not entirely clear, you can watch the demonstration film *The Helium Method* (Chabot 2012) that can be obtained from www.dyingathome.nl, in which an actor demonstrates all the steps in detail. These steps equally apply to nitrogen gas though an inert gas flow regulator is needed because of the high pressure in nitrogen tanks. In another film, *Eyewitnesses* (Chabot 2013) Els relates how she prepared the bag together with her seriously ill aunt, and witnessed her aunt dying by helium (See Intermezzo 2).

What do you need in case of helium gas?
- Two disposable helium tanks (14.9 cu. ft., weight 7 lbs.): one for several practice runs, both tanks (one partly full, the other completely full) for the final act. The tap on the second tank should not be opened before it is to be used.
- An 14.9 cu. ft. tank contains enough helium to fill about 50 balloons. The tanks can be purchased at shops that sell party goods, or on the Internet. Smaller tanks holding 4.5 cu. ft. are not recommended, because they might run out of gas too quickly.
- Soft plastic tubing measuring 3/8 inch internal diameter and 1/2 inch outer diameter: two pieces about 5 feet long. This can be found at hardware stores.
- A strong, transparent oven roasting bag measuring 19 x 23,5 inches or 48 x 60 cm. Do not substitute this with a less robust bag.
- An adjustable wrench.
- Additional items: a piece of elastic (about 10 inches long), a roll of tape, a pair of scissors and a large safety pin. That's all.

Figure 1.1 Putting the elastic in the hem of the bag and measuring the size of the neck

Figure 1.2 Soften the tube in hot water and push it over the nipple on the tank

What do you need in case of nitrogen gas?
- One nitrogen tank of 20 cu.ft.3 is enough to die. However I advise to buy two tanks: one for practice (see below why this is important). The other tank should not be opened, as it will be used on the day you actually end your life
- An inert gas flow meter allows the user to adjust the flow of gas to the desired 15 Liters/min. This device can be purchased in a welding shop or from a gas company over the Internet (for names see appendix 6)
- Soft plastic tubing that fits the outlet port of the tank. The diameter of the outlet port varies between some tanks. Take two pieces about 5 feet long. This can be found at hardware stores.
- additional items (oven bag, wrench etc) are the same as with helium.

Figure 2.1

Figure 2.2

Nitrogen tank from welding shop (right) and Max Dog Brewing tank (left) with tube. Chianti bottle in between the tanks

Nitrogen tank with inert gas flow meter set at a flow of 15 L/minute

How to prepare the bag and connect it to the two tanks
The procedure consists of several steps. First, place the bag flat on the table. To make a hem through which the elastic can later be threaded, fold the opening of the bag outwards and fasten it with short pieces of tape. Make sure that the hem is wide enough to thread the elastic through. Cut a small hole in the hem and use the safety pin to thread the elastic through the hem. Then remove the safety pin. For the time being, leave the ends of the elastic sticking out.

The opening of the bag must be made to measure. Put the bag over your head and pull it down over your chin. Then pull the two ends of the elastic until it fits your neck, but is not tight. It's a bit uncomfortable at first, but you can take your time, since there is plenty of oxygen in the bag. Grasp the ends of elastic under your chin where you want to tie them together, and remove the bag from your head. Then tie a double knot in the elastic and cut off the loose ends. The bag is now ready for use.

Punctures may occur during the above procedure, but you can easily test to see whether this is the case. Hold up the bag with one hand, closing it tightly at the opening, and then open the tap on the practice tank. It will take a strong hand to squeeze the bag closed, and it may be necessary to ask for help. When the bag is completely full, turn off the tap. Then hold the bag next to your cheek, to determine whether any helium is escaping. If so, prepare a new bag.

Now take the tanks out of their boxes. To expose the metal nipple of the helium tank, remove the plastic inflation nozzle, using an adjustable wrench. Then take one end of each tube and connect one tube to each helium tank. If your hands are not strong enough, ask someone to help you. To make the tube end pliable, dip it into hot water for a few seconds, so that it will fit snugly over the metal nipple on the tanks. Push each tube by twisting it until it presses right up against the metal of each tank. If the fit is not tight enough, use tubing with a smaller diameter. Then put the tanks back into their boxes, to prevent them from tipping over.

The next indispensable step is practice.

Why is practice necessary?

When the time comes, your mind will be elsewhere. The two actions that lead to death should be carried out in the right order: turning the tap and pulling the bag over your head. It is important to practice each one separately, until the procedure has become automatic. Always use the same tank for practice. The other one should not be opened, as it will be used on the day you actually end your life.

The first time you try to open the tap on the tank in order to practice, you will probably need an adjustable wrench. It is important that you are able to turn it on easily from a reclining armchair or from your bed. You must be propped up with cushions on both sides, to keep yourself from falling aside when you lose consciousness. If that happens, the bag will be torn from your head and you will wake up. Is the tap easy to reach? You should always use the hand over which you have most control. In case of helium, turn the tap only slightly, since otherwise it will empty too quickly, and you won't have a slow, steady flow of gas for at least 20 minutes. This is achieved by opening the tap only a little, less than about 1/8 of a full circle (which is the same as half of a right angle), as demonstrated in the film. In case of nitrogen, the flow regulator should be set to provide 15 Liters/minute of steady flow. Ask advice about this at the shop where you buy this device. Alternatively, during the first practice run you may need some help how to do this.

Before you proceed to put the bag on your head, remove earrings or other ornaments that might tear a hole in the bag. If you have long hair, tie it up in a bun so that all hair is well away from the neckline.

The second act that should be practiced is pulling the bag over your head, with the elastic running from your forehead to the hairline of your neck. Position the bag so that the tubes are at the back of your neck. Press the air out of the bag with one hand, and then pull the bag over your face, if possible with both hands. The bag will not be tight at the point where the tubes are attached, but that doesn't matter. In fact, the bag should not fit too tightly, since some space is needed to allow helium or nitrogen to escape and to push out the exhaled carbon dioxide. The flow of inert gas prevents outside air containing oxygen from entering.

I realize that all this sounds complicated, and after a first reading you may feel at a loss. This is why I made the film about the helium method. When someone demonstrates these steps in the proper order, all will be clear. If you get the impression that the tank you have been using for practice is almost empty, you might want to order a third tank. If you plan on practicing a lot, it is reassuring to have another tank available.

Before the planned date, you should prepare a letter in which you explain why this was the only option left. After your death, a written statement will be important for the authorities, who will want to make sure you weren't pressured into taking your own life. The more information it contains on your personal reasons, the more convincing you will be to outsiders. Include specific information such as, 'this is my decision,' 'I have carefully considered my options,' 'nobody has put any pressure on me to die,' 'I have carefully researched this method and I have not been influenced or pressured in any way'.

3.5
Implementation

Choosing the right moment is difficult; and as you prepare all the equipment for use, you may still be agonizing over such questions as: 'When the time comes, will the children accept my decision? And how can I be sure they won't run into trouble afterwards?' If these considerations are still bothering you, then you are not yet ready to proceed. A great deal of discussion is necessary before you leave this world. Dying well means not dying alone, and not burdening your loved ones with worries about suddenly finding you dead. It is up to you to decide when the time is right, but do not hasten the last decision you will ever take.

When the day comes, don't forget to give the tap on the full tank a turn with the wrench to loosen it. When you've said goodbye to your loved ones, place the bag on the top of your head, with the elastic running from forehead to neck. With one hand press the air out of the bag, and with the other hand open the tap (or taps in case of helium) just a little. The bag will slowly fill up with helium. If this takes between 30 and 60 seconds, the taps are open far enough. If it goes faster, it's advisable to close them a bit. Wait until the bag is completely full, and then pull it over your head until it is under your chin, as you have

TIMER

practiced.

Right up to the last moment you can still change your mind, even when the gas is flowing and the bag is already blown up. The final steps are up to you and you should feel free to postpone the procedure. But if the decision is taken in time, with all steps carefully rehearsed, you will be able to end your life without assistance.

If these steps are impossible, because your illness has progressed too far, you can ask a compassionate person to risk criminal proceedings in order to help you. That is admittedly an awful thing to ask. In the case of a progressive illness, such as motor neuron disease or Alzheimer's, it is best not to postpone your last goodbyes too long, and to take responsibility for your own death.

One important thing you can do for your loved ones is to protect them against the law. The last thing you want is to get them into trouble. In order to prevent that from happening, you must turn the tap on and pull the hood over your face without any assistance from anyone else. Say clearly, 'Now I want you to film what I am about to do, on your cell phone or video camera, to prove that I did it myself'. It is wise to use a fresh camera and only one take. Do not redo the talk or delete anything, because the police can recover that anyway.

What do eyewitnesses see?

The information in this section is collected on helium deaths. It should be available soon about nitrogen deaths.

There is no choking. After about five breaths, the eyes may fix open with an astonished look, which is followed instantly by loss of consciousness. The eyes may turn upwards, the neck may relax and the head may fall to one side. Breathing accelerates in the first minute or so, but soon slows and become shallower. Sometimes there are short gasps that may be a few seconds apart, or up to a minute apart. The pulse will continue to beat for some five to ten minutes. There may be slow contraction or extension reflexes of the arms or legs, or both: the tongue protrudes or an arm or leg is extended. This may be a shock for witnesses who were not aware that these so-called 'stretch reflexes' precede the moment of death and are not consciously experienced. When you see them for the first time, it is as if the patient is regaining consciousness, but that is not the case. Sometimes there may be no breathing for a while, and then suddenly you hear a deep sigh or snore. This can continue for ten minutes or more, which in this situation is a very long time. The inert gas streams softly out of the tank, but after a while you no longer hear it.

When breathing has stopped, postpone removing the bag for at least fifteen minutes. This is a precaution to ensure that the person has passed away. After that, the hood may be raised in order to close the eyes. However, this may alarm the police. Be aware that after closing the eyes, it is normal for them to re-open on their own.

What to do with the inert gas equipment after the person's death
In many countries, those present at the scene of a self-chosen death risk being accused of homicide. Ogden reported that in many cases the inert gas equipment was removed prior to reporting of death.[9] Cause of death in such circumstances is usually attributed to an underlying illness. I do not recommend that one remove evidence of a self-chosen death.

In the tolerant legal climate of the Netherlands, those present usually leave the helium equipment untouched until the family doctor arrives and passes the medical file on to the coroner. The authorities read the farewell letter and watch the film on the cell phone, before departing with the helium equipment. The authorities usually reach the conclusion that relatives or friends were present to provide moral support by not allowing their loved one die in solitude.

If you decide to remove the helium equipment after the death of a very old or very ill person, a doctor will not usually suspect that helium gas was the cause of death. It will then be attributed to an underlying illness. Since this is deemed to be a 'natural death', the body may then be cremated or buried.

However, special chemical analyses have been developed for a post-mortem investigation to demonstrate the presence of helium and nitrogen in the blood or body tissues.[10] If one removes the helium equipment and the authorities become suspicious, the true cause of death may well be discovered.

3.6
Modifications

The inert gas method I have described applies the KISS principle (*Keep It Simple, Stupid*). However, others recommend the use of a pressure gauge or other devices. These and other modifications should be avoided in case of helium gas for the reasons set out below. Moreover, a handy person who is willing to assist risks becoming involved in a criminal proceeding.

A large helium tank

In some party shops it is possible to rent a large helium tank, which contains three times as much helium as the 14.9 cu ft. tanks. That sounds like an advantage, but there are also disadvantages. When you rent the tank, you will be asked to fill in your name and address, and it will have to be returned afterwards. After you have experimented with the helium for a while, there is no way of knowing how much is left. For these reasons we recommend using two disposable tanks with a capacity of 14.9 cu ft., one full and the other reserved for practice.

In case of nitrogen gas one tank will be enough for 30 minutes steady flow of gas.

Sleeping pills

It is important to be alert and very self-aware when making one's final decisions. It is not advisable to take sleeping pills before the tap on the tank is opened. Inert gases act so quickly that the pills cannot possibly take effect in shortening the time to death. If someone were to say a few last words between taking the pills and opening the taps on the tank, he or she might become drowsy, which would interfere with the acts that result in death.

T-piece

Some people prefer using a T-piece (available in hardware shops) to connect the two plastic tubes on the helium tanks with the plastic bag. That way, there is only one tube in the bag, instead of two. However, this means that there is greater risk of leakage, since you have to slide the tubes over the three arms of the T-piece. Each link means a greater risk of leakage. You can prevent this by anchoring the tubes with a clip, but that demands a bit of skill. You will have to decide what is feasible in your situation.

In case of nitrogen gas one tank will be enough for 30 minutes steady flow of gas

An inert gas flow meter

The method with nitrogen gas requires the use of an inert gas flow regulator because of the high pressure in nitrogen tanks. One can buy this at a welding or hardware store. The description below applies to the helium tanks only.

Most helium tanks are set to a working pressure of about 260 psi (pounds per square inch). In nitrogen this is 1800 psi. An inert gas flow meter should confirm that the tank is full. If the gas flow stops after a few minutes, it is likely that the tank has been opened by someone as a test and not securely closed before it was stored. The inert gas then leaks away, and after a few weeks or months the tank is half empty. This is why it is advisable not to turn the tap on before use.

It is sometimes suggested that a meter set to a gas flow of 10 liters per minute should be attached to the tube on the tank. However, in view of the KISS principle, it is advisable not to use devices you are not familiar with. Using fewer pieces of equipment means fewer connections, and less chance of leaks. There is no need for a flow regulator, provided that the person who opens the tank practices opening and closing it as discussed above. If the tap is turned only 1/8 of a circle, as previously advised, a full tank will produce a flow for at least 20 minutes.

If someone is still conscious after about four minutes while the helium is flowing into the bag, then the best course is to remove the bag oneself and look for the source of the problem. Either there is a leak somewhere, or the tank was not full at the start, or the tank was diluted with 20 percent air.

An anesthesiology mask

The plastic bag is regarded by many as an eyesore that interferes with the contact between the patient and his or her loved ones. Critics call the helium method 'undignified'. Some people suggest using a mask that covers nose and mouth, as is standard in anesthesiology during narcosis. Others recommend the type of mask used to treat sleep apnea. However, all efforts to replace the plastic bag with a mask have proved ineffective, because oxygen leaks into the area between the mask and the cheek. How is that possible? After someone loses consciousness, the cheeks change shape. Cracks form through which oxygen is sucked in, together with the helium. As a result, death comes slowly (sometimes taking as long as 40 minutes), and there are often lengthy and violent muscular spasms in arms or legs.[11]

3.7
An experimental test of the KISS-principle

Is there any evidence that the KISS-performed helium method is effective? To test this assumption, I conducted – with the help of others – an experiment with a dummy. This enabled us to measure the oxygen (O_2) concentration in the bag as a function of the time elapsed and several other variables.Measurements were done on a human-sized dummy (fig. 3.1). A mechanic constructed an artificial lung by fixing a piston into a cylinder that enabled us to simulate inhalation and exhalation. The piston could be moved along a scale (fig. 3.2) in order to measure how much air was moved into and out of a large bag filled with helium over the model's head. This bag was fixed loosely under an elastic band around the neck. A tube (9 mm inner diameter) connected the bag to a helium tank (fig. 3.3).

The cylinder was connected to an oxygen sensor (Greisinger GOX100), which was fastened in front of the dummy's head on a tube that connected the cylinder with the dummy (fig. 3.1). The sensor has a measurement error of plus or minus one percentage point. Before each test, the sensor was calibrated for the O_2 concentration in the room. At the first exhalation of about 250 cc after the helium bag has been pulled down, oxygen from the artificial lung mixes with helium; with each following exhalation, there is still some oxygen coming out of the artificial lung. When the quantity of oxygen drops quickly, as is the case in our experiment (see table 3.1), there is a lag of a few seconds

Figure 3.1

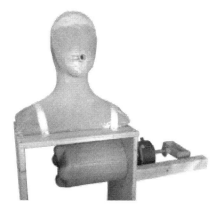

Dummy made of papier mâché with plastic mouth piece for oxygen sensor (fig 3); a tube runs from the cylinder underneath, at the back of the dummy, through its 'head' to the mouth piece.

Figure 3.2
Artificial lung

Cylinder with piston and scale that is calibrated for several equal parts of ± 250 cc air

Figure 3.3

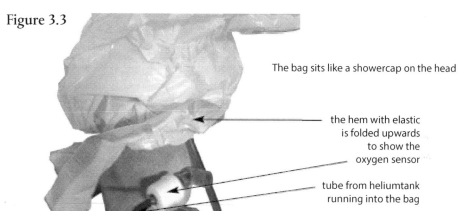

The bag sits like a showercap on the head

the hem with elastic is folded upwards to show the oxygen sensor

tube from heliumtank running into the bag

in the O_2 values detected by the sensor. All O_2 measurements provide relative values, i.e. they are indications of the O_2 percentage that comes out of the artificial lung relative to the O_2 percentage in the room in which the sensor is calibrated.

The first step in the experiment was to open the tap on the tank to fill the bag. We always used a bag with a size of 60 cm x 80 cm. At time zero, the bag was drawn over the dummy's head and sensor.

We tested the effect of four independent variables on the O_2 percentage in the bag (the dependent variable) over time:

- weak vs. strong flow, i.e. tap on tank < 1/8 (i.e. 45 degrees), open resp. 1/4 (i.e. 90 degrees)
- a small leak (several pinpoints in the bag) vs. a large leak (one finger)
- small and large gasps
- maximum exhalation prior to pulling the bag over the dummy

The experiment

The general reader may find this section overly-technical. For this reason, a summary is given in Textbox 3.3.

Textbox 3.3 Summary of the experiment

Relatively simple equipment – a dummy, an artificial lung and an oxygen
meter – was used to measure whether the oxygen concentration in an artificial lung goes down after a bag filled with pure helium is drawn over the dummy's head. At the first exhalation, oxygen from the artificial lung mixes with the helium in the bag. It is driven out of the bag along the neck by the steady flow of helium. This sequence continues during the subsequent exhalations which contain a decreasing quantity of oxygen.

The measurements in table 3.1 show:

(1) that a weak flow of pure helium from the tanks is as effective as a strong flow to bring the oxygen level down;

(2) that small pinpoint leaks in the bag have no significant effect on the oxygen level; a big leak raises the oxygen level.

(3) that large gasps have an effect for about 2 minutes, but do not raise the quantity of oxygen in the bag to a level where consciousness might return.

(4) Not shown in table 3.1: a maximum exhalation prior to pulling the bag down has no significant effect on the rate with which the O_2 percentage in the bag goes down.

In conclusion: although the experiment uses simple equipment, the results demonstrate that when a **kiss**-type set-up is used, there is a rapid fall in the oxygen percentage from 21% (in the artificial lung) to a level below 5%, which is incompatible with consciousness.

Table 3.1 shows the course of the O_2 percentage in the bag from the moment the bag is pulled over the face: oxygen that is exhaled from the lungs is mixed with the helium in the bag and pushed out of the bag by the steady flow of helium. The next exhalation of about 250 cc will contain both oxygen and helium that will be pushed out again. And so on, until after a few breaths, the lungs are almost completely filled with helium. Thus the O_2 concentration measured in front of the mouth of the dummy falls from about 21% in the room to below 5%. Consciousness is lost at an O_2 percentage of about 6-10%.

Table 3.1 O_2 percentage measured before and after the bag with helium is drawn down.

Test conditions: all tests with 14.9 cu. ft. helium tank and 60 cm x 80 cm bag. About five to six inhalations and exhalations of about 250 cc per minute is a normal rate and volume of respiration when a person is at rest.

	1	2	3	4
Test condition 1-4	Soft flow*	Loud flow*	Big leak when body tilted by 90 degrees	Small leaks (pinpoint)
O_2 % in room At time zero before bag over the face	20.9	20.9	20.9	20.9
O_2 in bag * After 30 sec (in/exhalation 250 cc each, 10-12 /min)	7.5	10.9	7.8	7.6
O_2 after 60 sec	3.9	6.0	4.5 large leak	3.7
O_2 after 90 sec	3.9	4.7	6.4	3.1
4 min gasp** in/ex 1000 cc	1.4 to 5.5	1.2 to 4.5	10.6 to 12.7	1.6 to 4.9
6 min gasp	1.6 to 5.2	1.4 to 5.8	14.1 to 14.4	1.7 to 6.6
10 min gasp	2.0 to 5.4	7.3 to 8.3	—	1.5 to 4.9

– soft flow: tap turned less than 45 degrees
– loud flow: tap turned 90 degrees
** deep inhalations and exhalations of about 1000 cc at 2-minute intervals

Findings

When the valve was opened 90 degrees (1/4 of a full circle), we measured the flow for 10 minutes. When it was opened 45 degrees (1/8 of a full circle), we stopped the test after 20 minutes. The tank was not yet empty.

In both cases (valve opened 1/4 or 1/8), the oxygen level at the mouth dropped rapidly (i.e. within two minutes) to 3% oxygen (note that the error of measurement is ± 1 percentage point; at low oxygen levels, the error of measurement might be larger).

Based on these measurements, we recommend that the valve should only be opened 1/8 of a full circle. This provides at least 20 minutes of helium flow and the O_2 percentages drop as quickly as is the case when the tap is opened further to 1/4 of a full circle. However, in this latter case, there is only ten minutes of flow, which may or may not be effective in causing death.

In test 3 we tested the effect of a big leak by tilting the dummy by 90 degrees after one minute. The effect is twofold: the tubes pull the edge of the bag up and the mouth becomes level with this opening. Thus air flows into the bag, as is shown by the rise of O_2 in test 3.

Test 4 shows that 3-4 pinpoint leaks in the bag have no perceptible influence on the rate with which the O_2 percentage in the bag goes down.

When someone is not properly propped up in bed with cushions, he or she will fall sideways in the same way as the dummy. He or she is unlikely to die, but will probably recover consciousness when the tank is empty. So far, no permanent brain damage has been reported to us or in the literature.

We also tested whether a deep exhalation before pulling the bag down over the dummy head would make a difference to the O_2 percentage (not shown in Table 3.1). We found no difference, perhaps because of the delay of the sensor. We would suggest that a deep exhalation may have some effect on the time to unconsciousness, but this is not proven. If in this stressful moment, the person forgets to exhale deeply, this should not be a cause for concern, as consciousness will soon be lost anyway.

The number of fingers space between the elastic band and the neck does not make much difference either, as long as this opening is far enough – at least four fingers – below the mouth. If it were level with the mouth, this would have the same deleterious effect as a big leak in the bag. It is really important to sit up straight and not lean sideways.

The function of the elastic band around the neck is not meant to keep O_2 out or helium inside, but to prevent the bag from being lifted upwards due to the rising effect of helium.

A striking phenomenon for those present at the scene of a helium death are the sudden deep inhalations and exhalations called 'gasps' that occur a few minutes after the person has lost consciousness. We were interested to find out what the effect of these so-called 'gasps' is on the O_2 percentage in the bag. One might expect a lot of O_2 to be sucked into the bag.

With the cylinder in the artificial lung, we could mimic a gasp by a 1000 cc-inhalation and exhalation. As shown in Table 3.1, we found a small but stable effect of gasps. The oxygen level immediately rises, but never reaches a level of more than 6% as long as there is no big leak. Over a period of two minutes the O_2 percentage returns to the previous low level. There are three equally plausible explanations for this gasp-effect on the O_2 percentage:

– The inhalation into the artificial lung contains the (low) oxygen level within the bag. The outward flow contains the remaining air (oxygen) from the artificial lung at the time the test was started.
– A gasp sucks in some air along the neck which contains 20% of O_2; after exhalation, this is pushed out of the bag by the steady flow of helium.
– The piston in the cylinder may have a small leak through which O_2 enters the bag; this is pushed out of the bag by the steady flow of helium.

Whatever the right explanation may be, and though they may be stressful for people present, such gasps will not interfere with the self-chosen death.

In conclusion, although the measurements have been made using relatively simple equipment, the results demonstrate that when a KISS-type set-up with helium gas is used, there is a rapid fall in the oxygen percentage from 21% (in the artificial lung) to a level below 5%. A 5% level is incompatible with consciousness.

3.8
A personal evaluation

Sooner or later, we all form a personal preference about how we would like to die, whether it is in our sleep from a heart attack, or from a stroke at a ripe old age. For those who want to be in control, the medication and the helium methods provide a way to do it in one's own time, in one's own bed, with someone at one's side.

Ever since the Second World War, the thought of dying 'from gas' has had a sinister ring to it, because for some it recalls the Holocaust. However, in countries where the law does not permit a physician-assisted death for terminally ill patients, many people are grateful that helium exists and can bring about a fast and painless death.

Another advantage of helium is that one loses consciousness quickly without suffocation and one's heart stops beating quickly, usually within ten minutes. Some people fear to lie for hours in a coma, sometimes incontinent and retching. They therefore opt for helium and not for a lethal drug.

A serious disadvantage of the helium method is its suspect image. If one decides to remove the equipment after death, this implies years of secrecy for those present. Leaving the equipment next to the body means a police investigation will follow. In a few countries, including the Netherlands, Switzerland[12] and Germany,[13] assisting someone with the preparations – making the hood and hooking up the tanks – is not a crime. The further ahead one plans, the less likely it will be that the purchase and hooking up of the tanks will be seen as a criminal offence. The police will probably want to know if the deceased ended his or her life voluntarily and after due consideration. This can be verified by letters from medical specialists that make mention of a terminal diagnosis, and by a farewell letter from the person concerned, giving their reasons for their choice for death. What the police really want to know is whether the actions that caused death were actually carried out by the person themselves. Tape recordings of conversations may help to convince the authorities, as well as a film that has been made of the implementation.

The investigation by the coroner can be extremely trying. But if the confrontation with the police is carefully planned, the letter drawn up by the deceased is clear, and the event has been filmed, it is unlikely that those present will be prosecuted. There may be exceptions to this, such as in Australia and other countries, which have repressive laws.

Some people object to the use of a transparent plastic bag, as it prevents a loved one from touching your face as you are dying. Your child will have to wait before closing your eyes. This cannot be helped. There is no such thing as an ideal method. We all have to weigh one option against the other and reach our own conclusion.

Chapter 4

The search for the holy grail

4.1
What would an ideal autonomous method look like?

Sociologists have found that there are a number of universal characteristics of a 'good', 'gentle' or 'peaceful' death: dying at the end of a full life, at home, surrounded by those dear to you and without violence.[1] In many cultures, a 'bad' death means dying alone or surrounded by strangers; in the prime of one's life; or in a violent way, by suicide or due to an accident.

In countries with good health care, great value is also attached to dying without pain and to being in control. Thanks to palliative care, this has become possible in many but by no means all cases. The quality of palliative care is still lacking in the USA,[2] for example, and many are unable to afford good treatment. It is thus unsurprising that demand for humane, self-directed methods of dying is increasing. In some countries and US states, under strict circumstances doctors are allowed to provide physician-assisted dying: one falls into a deep sleep or coma and respiration stops within fifteen minutes.

People everywhere ask themselves whether it is possible to die a 'good death' without the help of a doctor who provides lethal medication. In chapters 2 and 3, the answer was affirmative provided that the preparations are made that I have described there. However, people are searching for options that are simpler to obtain. There are serious problems with all of them and this is what we shall look at closely in this chapter. For instance, the success/failure rate without assistance by someone present has not been researched in some of these methods. In some other very effective methods their safety for those present has not been established.

4.2
The SESARID criteria

I have held over 50 in-depth interviews with relatives who had attended a self-chosen death with lethal medication. This enabled me to construct an online questionnaire that was sent to a representative sample of more than 21,500 Dutch citizens. After a strict selection process, 47 cases were identified that fulfilled my definition of a humane self-chosen death.[3] Seven criteria emerged for the ideal autonomous method, which together form the acronym SESARID (Textbox 4.1).

In Australia, Nitschke (2006) sent a questionnaire to members of his organization, Exit International. On the basis of the answers, he drew up a list of criteria that should be satisfied in the case of a peaceful death. Four of the characteristics in his Exit RP (i.e. Reliable and Peaceful) test are the same as the first four in text box 4.1. He does not mention number 6: 'no damage in case of failure of the attempt to die.[4]

Textbox 4.1 Characteristics of an ideal autonomous method: SESARID

1. **S** — **Safe** for others: relatives can be present without danger to their health.
2. **E** — **Effective**: almost always.*
3. **S** — **Sleeping**: death comes without pain or suffocation while asleep.
4. **A** — **Available**: the lethal medicines can be obtained with some effort.
5. **R** — **Responsibility**: The acts leading to death can be accomplished by yourself, without any help from others that would put the responsibility for your death on their shoulders.**
6. **I** — **Injury**: no damage in case of failure of the attempt to die.
7. **D** — **Detection-prevention**: difficult to detect in standard post-mortem examination.

* Even with 9 grams of barbiturates, it may take over 24 hours before cardiac and respiratory arrest occurs.[5] Death follows within 2-48 hours in over 99% of accurately reported cases.

** Eyewitnesses have reported that those present sometimes provide help to accomplish death. For instance, when using an inert gas method, turning the tap on the tank and pulling the bag over one's head. This is a criminal act, however compassionate one's motives. A video can show whether the deceased has performed the last acts that cause death himself.

Beside the similarities, there are also several important differences between the Exit RP test and SESARID. In part, these relate to differences in legislation between Australia and the Netherlands. In Australia, even staying at the self-chosen death of a close relative because one does not want to leave him or her to die alone may constitute a criminal offence.[6] There is a universal feeling one should stay with a dying person one loves. A lethal method that leaves no obvious trace after death (like with an inert gas) becomes the preferred choice under repressive legal circumstances.

The Netherlands has a long tradition of tolerance on moral issues, such as on abortion, gay marriage and the use of soft drugs, for example. Debating and being able to talk about moral issues is very important for the Dutch: family members and friends do not want to leave their loved ones to die alone, and increasing numbers of people are having the courage of their convictions. In Dutch culture, the most important thing is that the method is Safe for the loved-ones present. In the SESARID criteria, 'safe for others' therefore carries more weight than 'detection-prevention'. 'Responsibility' – taking responsibility for one's own death – is also considered important in the Netherlands. Someone who offers to help to perform the acts that cause death is breaking not only the law, but also a moral commandment ('thou shalt not kill') – even if this is at the other's urgent request. Taking responsibility for one's own death is missing from Nitschke's criteria. There may be good reasons for this, due to repressive legislation.

In writing this chapter I have consulted the Dutch pharmacists Paul Lebbink and Annemieke Horikx as well as the toxicologist Ed Pennings. Their advice was essential to sort out why the methods in this chapter do not fulfill some of the SESARID criteria.

4.3
Autonomous methods that do not meet the SESARID criteria

Textbox 4.2 Methods that fall short of SESARID in alphabetical order: Lethal gases, ligature method, medicines and poisons

1. lethal gases	– carbon dioxide (CO_2) is the gas that causes death in the plastic bag with sedatives method.
	– carbon monoxide (CO) is a lethal gas in exhaust of old cars or is produced by burning charcoal in badly ventilated spaces.
	– hydrogen sulfide gas (H_2S) is a lethal gas that has the smell of rotten eggs.

2. ligature or compression method

3. medicines	– anti-depressants (tricyclics):	amitriptyline and others
	– anti-diabetic medicine:	insulin
	– anti-epileptic medicine	phenobarbital
	– cardiac medicines:	digoxin and others
	– hydroxychloroquine	
	– opiates	morphine, methadone, heroine

4. poisons	
in nature	– animal
	– botanical
in chemistry	– cyanide (CN)
	– pesticides in agriculture
	– potassium chloride (KCL)

The methods in textbox 4.2 are being discussed in right-to-die circles. Some have a bad reputation, others a positive aura. I will discuss them within the framework provided by the SESARID criteria. This list is incomplete because new ideas will come up in the future.

4.3.1
Gases

In chapter three I have discussed helium and nitrogen gas for a self-chosen death. Both methods fulfill the SESARID criteria for an autonomous and humane death though in most cases they need some help with the preparations like carrying the tank into the room and connecting the tube with the tank. In the Netherlands these preparations are not considered a punishable offense provided the acts that cause death are done by the person who is going to die: turning the tap on the tank and pulling the bag over one's face. This has been discussed in the introductory chapter.

In this section I will discuss to what extent lethal gases that are well known in the right-to-die literature fulfill the SESARID criteria: carbon dioxide (or CO2) carbon monoxide (or CO), and hydrogen sulfide (H2S). Gunnell (2015) has collected unique data on the frequency with which these methods have been registered in post mortem examinations in England and Wales (see appendix 4).

Carbon dioxide (CO2)
co2 is the cause of death in the plastic bag with sedatives method.

The use of a plastic bag as a supposedly effective suicide method was made known in 1991 by Derek Humphry in his book Final Exit. A plastic bag is placed over the head, made more or less airtight with elastic bands around the throat, and used with sleeping pills. The advice is to keep the bag open with one's thumb until one falls asleep, before the exhaled carbon dioxide can produce symptoms like nausea, headache and suffocation.

Carbon dioxide in a high concentration of about 4% or more will build up in the bag and causes death. It is often erroneously assumed that the lack of oxygen in the bag causes death. In theory death comes when the person has fallen into a deep sleep by the sedatives. According to Derek Humphry, if properly carried out in this way "It is foolproof."[7]

Between 1994 and 2006 the opposite conclusion was reached by four researchers independently from each other.[8] They reported that the method failed to be effective for different reasons. For instance, it is well known that the rise of carbon dioxide (CO_2) inside the bag that is exhaled from the lungs gives rise to frantic movements. Thus, in some cases those present provided assistance in suicide by replacing a displaced bag or by holding down the person's hands when he or she is sedated and wrestling for air. Of six cases that were reported to me in detail by outsiders who had been present and had not provided assistance only two cases died.[9] Due to an overdose of medicines others died after several hours when the bag could not have caused death anymore. An unknown number wakes up and does not report their failure.

It can be a traumatic psychological experience for the person who wakes up after a rational decision to end his life.[10] When it succeeds it can be traumatic for the attendant relative or friend who has assisted in some way. Already in 1994 John Pridonoff, an executive of the Hemlock society, wrote: "Even when it works, the person who assures its success must live out the rest of his/her life with graphic dreadful memories of the event, while facing possible criminal liability. The dying individual who seeks to apply it without the help of another faces the likely prospect of failure."[11]

Docker has argued that the method can be made more effective by using a very large plastic bag and testing the dosage of sedatives for a deep sleep beforehand.[12] However, no detailed observations have been published in cases where these precautions had been observed. We simply lack research data by independent observers on the rate of successes and failures.

Another expert has abandoned this method and crossed over to inert gases: "The best method of using an Exit Bag involves the use of an inert gas, such as Nitrogen, Helium or Argon."[13]

As long as experts do not reach consensus what the proper modifications of Humphry's method should be, any empirical study will remain inconclusive. Lack of research that provides observational data, for instance on video, stands in the way of clearing the fog around this method.

Recently helium tanks (often sold for filling balloons) have been diluted with oxygen. This has given rise to discussion among experts about an alternative for the helium method.[14] Derek Humphry and

Chris Docker have suggested that the plastic bag with sedatives might provide an alternative when the helium method will not be available anymore. However, it is my impression that other professionals do not take the plastic bag with sedatives method serious as an alternative for inert gases.

A discussion of CO_2 as a lethal gas would not be complete without recalling how it has been used in the Netherlands during an outbreak of highly contagious avian flu, which threatened the livestock. CO_2 gas was discharged into battery cages containing thousands of chickens, and proved an effective means of killing all the animals by suffocation. For private individuals it is difficult get hold of tanks with CO_2 gas.

Carbon monoxide (CO)

Carbon monoxide is not an inert gas, such as nitrogen or helium gas. It binds itself to hemoglobin, preventing the blood from transporting sufficient oxygen to the brain. Strictly speaking, it is not a chemical bond between CO and hemoglobin, but a much stronger Van der Waals force. This makes CO such a dangerous gas for those present who inhale just a little bit of it that may accumulate and cause some cerebral damage. Thus, this procedure must always be carried out alone.

Before the 1990's, car exhaust fumes contained carbon monoxide. For many years, this was a regular method of suicide.[15] Exhaust fumes were passed back into a sealed car via a hose. The smell of the fumes is very unpleasant. After some time, the person would fall asleep and death would follow, if the motor did not cut out. In modern cars since the nineties, carbon monoxide is caught and removed by catalytic converters. As a result, there is now so little carbon monoxide in exhaust fumes that it can take hours for death to occur. If the attempt fails because the motor cuts out or the person is found while still alive, there is a significant chance of brain damage. The I in SESARID is not satisfied.

Carbon monoxide is also produced when charcoal is burned in an enclosed space, because there is too little oxygen for full combustion. This can also occur, for example, with the burning of butane or propane gas from a camping stove. When there is little oxygen, the color of the flame changes from blue to yellow. Whether from charcoal or a camping stove it takes a number of hours for a person to die, and other people in the same or an adjoining room are seriously at risk. This method fails the first S of SESARID

Neal Nicol has described how Jack Kevorkian, in a pick-up truck, would use a hose to pass carbon monoxide gas from cylinders to a mask over the patient's nose and mouth. A seriously ill person would die quickly, after which Kevorkian immediately closed the tap on the cylinder to prevent harm for those present in the room.[16] For laypeople cylinders containing carbon monoxide are very difficult to get hold of. The A in SESARID is not fulfilled.

Nitschke has described how a chemical reaction between two acids can produce carbon monoxide (CO), which can be passed to someone's nose by means of a plastic tube. He calls this apparatus the 'CO-gen', and argues that retired elderly people who wish to have a means of dying a gentle death can make such an apparatus themselves, using materials that can be bought from supermarkets and hardware stores.[17] There are no eyewitness reports of people who have built this apparatus and who have then used it to die. Eyewitnesses might run the risk of brain damage. Dying by CO gas from charcoal or exhaust gas remains a lonely method and carries a serious risk of brain damage for the patient, if insufficient gas is produced and death does not follow. It is not safe and it is not reliably effective.

Hydrogen sulfide (H2S)

Hydrogen sulfide is sometimes referred to as 'rotten egg gas' because of its awful smell. Its use in suicide was first noted in 2008 in Japan in a pact suicide (appendix 4). When a chemical that contains polysulfide is mixed with an acid, hydrogen sulfide gas is produced. Inhaling this gas at a concentration of 1:1000 is reported to be instantly lethal. It kills as quickly as cyanide (see below), possibly because it poisons cellular metabolism.

Gerald Metz is researching whether and how this method can be made comfortable and safe for those present at the deathbed.[18] He reports an experiment with a mouse that, just before its death, showed no tendency to avoid or escape the environment. No nausea, vomiting, or seizures were evident. Experiments with larger animals are in order to confirm the lack of uncomfortable side-effects.

According to Metz, the best source to produce H2S gas is an insecticide. Its active ingredient is calcium polysulfide. Mixed with almost any acid it promptly begins to generate significant quantities of H2S. This is an uncontrolled chemical process that can go on for a long time. Professionals have exchanged views on how to turn this into a controlled chemical reaction. So far without any results.

Gunnell (2015) has reported 14 H2S suicides of which two cases were accidental deaths. A serious drawback of hydrogen sulfide gas is that casualties have been reported when the gas leaked into adjacent rooms or rescue personnel were overcome when they rushed into the place before they could escape.

4.3.2.
Ligature or compression method

The essence of this method is to block blood flow to the brain by compressing the carotid arteries in the neck. A tourniquet is placed around the neck, above the Adam's apple. The windpipe should stay open so one can freely breathe. If there is any discomfort this can be prevented by putting a pad between the windpipe and the device. The person choosing this method tightens the tourniquet with a long-handed spoon, wood stick, metal rod or other device. When tightened by turning the spoon, one will become dizzy. With a few other twists, fainting will occur. The person should be lying on his or her side, inclined downwards, so that the tourniquet won't unwind. According to Docker death will come in 5-10 minutes.[19]

Some have pointed out that the brain still gets blood through arteries protected by the vertebrae and this might be sufficient to keep the heart and respiration going. However, even if the brainstem receives blood, death will still occur due to swelling and hemorrhaging inside the brain from lack of blood via the carotid arteries.

In 2012, Neal Nicol introduced the inflatable neck collar as an alternative technique for the compression method. When inflated, the collar puts pressure on the carotid arteries only. With his permission, I include a description of how he made the collar (see textbox 4.3).

Textbox 4.3 How to make an inflatable neck collar for the compression method

The collar is a 4″ PVC pipe connector cut in half. I inserted two pediatric blood pressure bladders to correspond with the position of the carotids with a blood pressure pump that is used to inflate the bladders with a Y connector to deliver air to both bladders. I connected the collar with a Velcro hook and eye adhesive strips that can be adjusted for the size of the neck. When the collar is fitted to the neck size and the bladders positioned over the carotids, the bladders can be pumped up and sealed with the hand bulb to maintain pressure. There is no pressure on the windpipe because it sits between the two bladders.

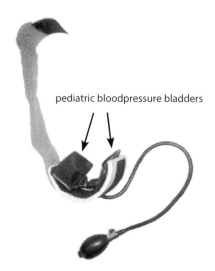

pediatric bloodpressure bladders

Printed by courtesy of Neal Nicol

There have been no reports of a self-chosen death by the compression method that was attended by relatives or friends. It appears to be an lonely method. Not necessarily so, but those present would definitely be suspected of homicide. A video and other precautions showing that no assistance was given in performing the lethal acts would be necessary and might give at least some protection.

Video recordings have convinced a Dutch higher court that that no punishable assistance had been given by a son who recorded how his 99-year mother took a lethal cocktail of chloroquine and valium he had given her.[20] The Farewell Foundation in Canada has advised some who want to be present at a self-chosen death with helium to film it on a cell phone.[21] Whatever the precautions, harsh treatment by the authorities is always a possibility.

Practice beforehand with the compression method would be necessary as problems may arise with, for instance, the position in bed or how to turn the spoon till dizziness comes and how to prevent the spoon to unwind. At this crucial moment many persons will lose the ability to turn the spoon fast enough while preventing that it unwinds because one is partly unconscious. This is not only hard but even dangerous to practice beforehand.

Someone present may be seduced to assist by giving a few more twists. Like in the plastic bag or helium method people have assisted in the final acts to assure that their loved one will not awake. In most countries, such assistance is a serious crime, even though suicide itself is not.

If a relative is brave enough to be present at this method he should be aware of the side effects. Like in suicides by hanging, someone who dies in this way in bed will be likely kicking the legs, which may be disturbing to witness. He or she will also be soiling himself.

A particular pressing problem with the compression method is that there are no data on the rate of successes and failures when it is practiced alone. No eyewitness reports by independent observers have been published. With the helium method this is different. I received first hand eyewitness reports from two physicians (the one American, the other German) who had been present at a large number of persons who had died with helium. Their reports fully concurred.

What is more, criminologist Russel Ogden recorded several helium deaths on video. He also analyzed several cases that were recorded on video by the Swiss organization Dignitas.[22]

All this systematic evidence that has been collected for helium gas is not available for the compression method. The lack of recorded cases and of eyewitness reports, either in print or by several different experts, resembles the present state of knowledge regarding the plastic bag method. Data are still far from complete. The compression method is therefore still in an experimental phase.

For progress to be made, it is essential that the pros and cons of this method should be discussed by a group of experts. In my opinion concurring and dissenting opinions should be published for laypersons.

4.3.3
Medicines

Anti-depressants with a tricyclic chemical structure
According to the toxicologist Pennings, in appropriate doses these drugs are very likely to cause death by cardiac arrest, and possibly by depression of respiration (apnea) as well.[23] Amitriptyline will be discussed here (and not in chapter 2) because less than 20 cases have been reported by experts in which it was used for a self-chosen and humane death.

Availability: tricyclic antidepressants do not fall under international drug control conventions and thus are usually easier to obtain than opiates and barbiturates. See chapter 1, section 1.3 for more information.

In the European Union a prescription is required for these drugs, but not all (Internet) pharmacies live up to this requirement, as tourists in Southern or Eastern Europe have found out.

Like chloroquine, tricyclic antidepressants can be ordered over the Internet or without a prescription from pharmacies in many countries outside the EU and the USA.

Table 4.1 Amitriptyline

Brand names	Tryptizol, Amitriptyline, Elavil, Endep, Amitid, Amitril.
Availability	From Internet pharmacies, or without a prescription in many countries outside the EU and the USA.
Prescribed	Primarily in cases of depression, but also for intractable sleeplessness or chronic pain.
Habituation	None. It is recommended that one is 'clean' from sleeping pills for two weeks.
Lethal dose	7.5 g, e.g. 150 tablets of 50 mg.
Sleeping pills	It is essential to take 500 mg of long-acting benzodiazepines (BDs).
How to take	Pills should be ground and ingested together with the ground sleeping pills. The powder can be sprinkled over yoghurt. Then drink a glass of water, milk or juice.
Time to death	12-48 hours, as antidepressants are absorbed slowly.
Reported cases	Psychiatric patients have been known to hoard tricyclics without knowing the lethal dose, meaning that some died and others failed. Other antidepressants are not lethal.
Side effects	Some patients run a high temperature, which should not distress those present. Any epileptic seizure will be suppressed by diazepam.

Several other tricyclic antidepressants are prescribed for depression. All the information is the same as for amitriptyline. In alphabetical order (brand names in brackets):
 – desipramine (Pertofran, Norpramin)
 – dosulepin (Prothiaden)
 – doxepine (Sinequan, Adapin)
 – imipramine (Tofranil)
 – nortriptyline (Nortrilen, Aventyl, Pamelor)
 – trimipramine (Apo-Trimip, Surmontil)

Anti-diabetic medicine

Insulin: one occasionally reads in the newspaper that a doctor or nurse has killed a dementia patient with insulin. An overdose of insulin can *sometimes* be lethal for elderly people who are in a bad physical state. The toxicological literature shows that the lethal effect is very uncertain for people who are not seriously ill. As failed attempts using insulin do not make it into the newspapers, this creates the incorrect impression that insulin is always lethal.

Anti-epileptic medicine

Phenobarbital: the effectiveness of this drug has not been proven yet. It is discussed in this chapter (instead of in chapter 2.1) because less than 20 cases have been reported in which it was used for a self-chosen and humane death.

Table 4.2 Phenobarbital

Brand names	Luminal, Gardenal, Phenobarbital, Phenobarbitone.
Availability	Prescribed only for epileptic patients who do not respond to other anti-epileptic medication.
Sleeping pills	Long-acting benzodiazepines are necessary.
Lethal dose	6 g, i.e. 120 tablets of 50 mg or 240 tablets of 25 mg.
How to take it	Oral route. Grind the tablets together with the sleeping pills. Sprinkle over custard or yoghurt, stir, and eat in silence.
Time until death	From 12 to 72 hours.
Anti-emetics	Strongly advised.
Reported cases	Very few.
Habituation	If used on a regular basis, habituation occurs. One should be 'clean' for at least 4 weeks.
Disadvantages	1.Very slow-acting barbiturate. 2. Difficult to collect from epileptic patients who might use less than the prescribed dose, which for them would be dangerous, as this could trigger a seizure.

Cardiac medicines

Digoxin: some years ago, I asked an internist which medicine would be suitable for use in overdose in a humane self-chosen death. He answered, 'I would use a handful of digoxin (Lanoxin): the body goes into shock, just as with a heart attack, and death comes quickly'. However, research into a large number of digoxin poisonings, which occurred when a factory erroneously put too large a dose in the pills, revealed that only 7% of the heart patients died. They only died when they had been taking the overdose for a number of weeks.[24]

Verapamil: a different medical specialist was convinced that an overdose of verapamil (Isoptin), a calcium antagonist, would be lethal. Verapamil is a substance that slows down electrical conduction in the heart muscle. However, an elderly lady who was prescribed the drug, saved it and took it all at once did not even lose consciousness.

Doctors may declare with great certainty that a certain medication is lethal, even though they may only have witnessed a few cases. Of course, medicines that are taken in overdose can sometimes be lethal; in theory, even a kilo of table salt (NaCl) can be lethal. However, research in the toxicological literature shows that the medicines mentioned above, along with others, are unsuitable for causing death in most people.

Hydroxychloroquine

Hydroxychloroquine is prescibed for rheumatic pains. The effectiveness of this drug has not been proven yet. It is discussed in this chapter (instead of in chapter 2.3) because less than 20 cases have been reported in which it was used for a self-chosen and humane death.

Table 4.3 Hydroxychloroquine

Brand name	Plaquenil.
Prescribed	For rheumatoid arthritis.
Lethal dose	15 g, i.e. 75 tablets of 200 mg.
Reported cases	Very few. One failure: due to diarrhea, only part of the dose entered the body.

All other information is the same as in Table 2.7 for chloroquine phosphate.

Opiates: Morphine and street drugs

Morphine: All the information in the table on oxycodone (Chapter 2.2) in theory also applies to morphine. However few reliable and detailed reports by experts are available.

Brand or trade names of morphine: Morphine or Morphin, Skenan, Oramorph. 'Controlled release' morphine such as MS Contin CR and Kapanol CR enters the blood gradually and thus may not cause apnea. The pharmacists advise against the use of controlled release (CR) opiates.

Methadone: street drugs such as methadone are usually cut and often polluted.[25] However, in some American states methadone is part of the therapy offered to addicts. In that case, the dosage is known.

Table 4.4 Methadone

Brand or trade names	Symoron, Methadone, Dolophine, Mehadose, Matadol.
Prescribed	In some countries, usually for heroin or opiate addiction. From Internet pharmacies.
Availability	On the black market, provided one has a reliable middleman who can purchase it. Tablets contain e.g. 5 mg. The liquid is not lethal.
Reported cases	About 20.
Disadvantages	Difficult to get hold of and very difficult to kick habit when prescribed as a painkiller.

All other information in Chapter 2.2 on oxycodone also applies to methadone.

Diamorphine or heroin: in the UK, diamorphine, like morphine, is used as a painkiller. Taken orally, a lethal dose would be about 600 mg for someone who is 'clean'. On the black market, heroin is cut, which makes it unreliable for use as a lethal drug.

4.3.4
Poisons

Poisons in nature: animal or botanical

Time and again, one hears of someone who has died after ingesting a botanical poison or an animal venom. Indeed, some plants and toadstools contain extremely poisonous substances. The venom of some snakes, spiders and fish is also deadly. This is due to the presence of substances in the poison that inhibit the enzyme cholinesterase, or that contain a substance with a cholinergic effect.[26] These substances cause muscular paralysis, followed by a slow death through apnea: the person suffocates while fully conscious. Similar chemical substances are also used in poison gases.

One well-known example of a deadly botanical poison is curare, which is used as an arrow poison in the Amazon region. This poison paralyses the muscles and breathing. When injected into a vein, curare works rapidly to cause apnea while the person is still conscious. Other deadly botanical substances have less rapid effects, such as the poison that Socrates had to drink because he had been condemned to death. Some only lead to death after a few days or weeks, and are accompanied by intense pain.

The concentration of the poison differs for each plant or animal. Researching whether a particular botanical extract contains enough deadly poison, requires complex chemical determinations.

It is clear that poisonous substances that are to be found in nature are unsuitable for a last-will-pill: one can never be sure that one has collected a lethal dose, and the death that they bring is a painful one.

poisons in chemistry

Cyanide (CN)

If a person heats sulfuric acid in pan, sprinkles potassium cyanide crystals into it and then takes a few deep breaths over the pan, he or she will die of suffocation. While it is possible to acquire the two basic substances with some effort, the method poses a great danger to others in the room, who will breathe in some of the cyanide fumes. Cyanide is an extremely effective killer. Cyanide inhibits the ability of cells to use oxygen, not the transport of oxygen by hemoglobin.

Specifically, cyanide inactivates mitochondrial cytochrome oxidase, and that inhibits cellular respiration.[27]

In films about the Second World War, one often comes across characters who use cyanide to kill themselves quickly before falling into enemy hands. Nazi officers were said to have carried cyanide capsules. Biting on a capsule would release a cyanide compound that, when inhaled, would lead – more or less rapidly – to death.

Humphry cites conflicting eyewitness reports that describe the symptoms of a death by cyanide. Nitschke joins Humphry in his conclusion that, 'the balance of evidence about using cyanide indicates that it is best not used'.

Both authors suggest that there is an 'enigma' surrounding death by cyanide. According to the literature however, the cause of death is quite clear. In the few minutes that the brain can still function after cellular respiration is inhibited, the person may well experience death throes while conscious. If the person has inhaled a smaller quantity, this process can take longer.

Pesticides used in agriculture

There are hundreds of types of agricultural pesticides in use, and these are sometimes used in suicides. One well-known example is rat poison, of which there are a number of variants on the market. None of these poisonous substances offer a humane death; they take a long time and are accompanied by extremely painful symptoms.

The pesticides designed to kill rats and insects that are available in supermarkets have been chemically altered so that they are rarely lethal for humans. They do cause painful cramps and other nasty symptoms.

Potassium chloride (KCl)

If injected into a vein or into the heart, potassium chloride causes cardiac arrest. The person may experience pain in the few minutes before they become unconscious, due to ischemia in the heart muscle. Few fatal intoxications have been reported in the toxicological literature.

Potassium chloride can be found in the spices aisle in supermarkets, under the name 'NoSalt'. Taken orally or in a suppository, it causes diarrhea. As a result, most of it will never enter the blood. Moreover, the kidneys eliminate potassium chloride very quickly. It is therefore unlikely that, if taken orally or as a suppository, the potassium chloride will reach a lethal concentration in the blood. It might become lethal in the case of renal failure, but the dose that should be taken is uncertain.

Nicotine

Six hundred mg of nicotine is probably lethal for most human beings. A concentrated nicotine solution can be made in every kitchen. Nicotine from tobacco and e-cigarettes can be dissolved in 80% pure alcohol. The container for an e-cigarette contains a maximum of 36 mg of nicotine, and tends to contain less.

After drinking the solution, which tastes extremely unpleasant, a person will first suffer stomach cramps, nausea, vomiting and diarrhea. Within a few hours, the heart rate will slow and paralysis of the respiratory muscles will occur, which will finally end in death. There are no reports in the toxicological literature on how long it takes a person to die, but death will probably occur after about six hours.

If one were to take a long-acting and a fast-acting sleeping pill, as described in chapter 2 in the section on chloroquine, this would rapidly put one into a long-lasting deep sleep. One might be unconscious before the onset of the stomach cramps and the heart and lung failure. One would also need an anti-nausea drug, because the taste is so unpleasant that the solution may be vomited up when one is becoming drowsy.

I do not recommend this method, because there are no reports of it ever having been used. Therefore, there is no certainty that a concentrated nicotine solution can be safely ingested without vomiting and will effectively cause death.

4.4
Towards a cooperative culture of debate

All methods described in chapter 4 have been mentioned at one time or another by right-to-die experts. In my opinion none of them provides at present a serious alternative for the methods in chapter 2 and 3.

One obstacle to productive discussion among right-to-die experts is that they do not research what most people regard as 'dying well'. Therefore, no consensus will come within sight, what the SESARID criteria for dying well should be.

Observers should be welcomed at a self-chosen and dignified death provided they stick to a research role. First, they should never intervene and, second, they should share their observations with the police. Russel Ogden has set a standard for this as an observer of self-chosen helium deaths in Canada where the law seems less repressive than in other English speaking countries.

The lack of productive exchange of opinions makes it unlikely that consensus regarding methods for a self-chosen and humane death will be reached in the next decade. Nevertheless the right-to-die literature has made some progress since its beginnings 25 years ago by paying more attention to observations of self-chosen deaths. Specialists from other disciplines are being drawn to this field and their advice is, sometimes, taken heed of. Small steps towards consensus are definitely within reach. For instance, information on how to synthesize a lethal drug from freely available chemical substances can be shared with chemists and pharmacists. The same applies to tests for the purity of barbiturates that have been ordered through the internet.

Only if a culture of open debate arises among experts will it be possible to exchange and test opinions. Every two years a summary of the outcome might be published for laypeople. A constructive spirit of criticism and openness is of benefit to both professionals and laypeople in their search for a gentle death.

Appendix 1

Refusal of treatment and legal representative

Refusal of Treatment [1]

I, _____ ,

being of sound mind, have made the conscious and considered choice to stop ALL treatment, except palliative care, and to die. This is my legal right. I have made this choice since my pain/suffering/misery/disability have made my level of suffering unacceptable. I understand that there are alternatives to dealing with my condition that may prolong my living. I have considered them, and reject them all.

My agent or legal representative, whom I have already designated as having a durable power of attorney for health care, has been instructed to refuse my hospitalization under any circumstances.

My agent has also been instructed to refuse all treatments to prolong my life, such as cardio-vascular resuscitation, artificial respiration, artificial hydration and nutrition, and all other treatments doctors may want to start with.

Signed _____   Date _____

Print Name _____

A Warning: There are family members who are unwilling to allow a dying relative to die naturally. They can't let go, making decisions not for themselves but for the person who has a terminal disease, functioning *as if they were the proxy*.

"Let's do everything for dad," even if this is not what dad wanted. Dad can often be coerced – especially in his current debilitated state – into agreeing to something he would not want under normal circumstances for fear of being disappointing or because he's just too tired to disagree.

Having a detailed advance directive for health care with specific decisions about refusal of intubation, feeding tubes, dialysis, and antibiotics can help immeasurably but only if the surrogate, the proxy, the person who has durable power of attorney for health care has the backbone to refuse medication – including chemotherapy – and invasive tests often recommended by the medical providers. Unfortunately, the default position of the proxy making the full code decision is likely to result in aggressive treatments and a prolonged death with its accompanying pain and suffering.

Appendix 2

The Witness Role in Intentional Death

Printed by courtesy of Sid Adelman

The role of the family member or friend who will be in attendance when a person chooses to die intentionally needs some clarification. People who *have a terminal disease and are in hospice* care are expected to die. It's not a surprise to anyone, not to friends and family and not to the supporting hospice organization. There is almost never an autopsy or coroner inquiry when a person has been in hospice care.

If a person chooses an intentional death, it could be interpreted as suicide and assisting in a suicide is illegal and so no friend or family member would want to be assisting. The words you use for yourself and others are very important and so **you will never be assisting, aiding, advising, encouraging, abetting, helping, or supporting.** Neither will you be providing the means for the intentional death. Your role, if you choose to be there, will be to say goodbye. You will be a **witness, an observer;** you are there to honor the passing, respecting the life of the person, or solemnizing the death.

There are people who do not believe a person has the right to choose the time, place and method of their passing. That person might be a paid caregiver, a friend, a family member or even a hospice nurse. There is no reason for them to know how your friend or family member died and you don't have to answer every question you are asked. What should you honestly say when you are asked and what should you share?
- "He passed away peacefully." *because that's what happened.*
- "She died in her sleep." *because she fell asleep and died.*
- "His suffering is over." *because it's over*
- "I'm really going to miss her." *because you will*

The person who chooses an intentional death will be in full control and will be doing everything for him or herself. You will only be a caring and loving witness or observer, nothing else.

If you are the one to report the death, you will call hospice and tell them that you think your (friend, spouse, partner, father, mother,...) has died and ask, "What should I do?"

Appendix 3

A self-chosen death with barbiturates in Ireland

Taken from the Belfast Telegraph, 24 April 2015

In the first prosecution of its kind on assisted suicide in Ireland, Gail O'Rorke, 43, from Kilclare Gardens in Tallaght, Dublin, is charged with attempting to aid and abet a friend's death by arranging travel to the Dignitas clinic in Switzerland between March 10 and April 20 2011.

Bernadette Forde, 51, on June 6 2011 was found dead in a wheelchair in her living room having taken a lethal dose of barbiturates sourced from Mexico. She was unable to travel to Zurich after a travel agent alerted the police to the plan.

Ms Forde was diagnosed with primary progressive MS in 2001 and had to give up her job in the human resources department in Guinness. In 2008 she was confined to a wheelchair after a car accident in a car park after her leg seized on the accelerator slamming the vehicle into a wall. She spent four months in hospital after that, had multiple liver surgery and both knees were left shattered. Ms Forde received no at-home state healthcare other than a nurse visiting her six times during her illness.

The court heard O'Rorke went from being a cleaner to becoming her friend and carer over 10 years. She washed Ms Forde's feet and body, and despite the indignity, took her out in her wheelchair and responded to calls day and night. The court also heard O'Rorke, who received 30% of Ms Forde's estate in a will, was in a hotel in Kilkenny on the night of her death.

Judge Patrick McCartan directed the verdicts at Dublin Circuit Criminal Court over the alleged assisted suicide. O'Rorke pleaded not guilty to three counts in connection with the death. She was cleared of aiding and abetting Ms Forde's suicide between April 20 and June 6 2011 by helping her to procure and administer a toxic substance.

She was also found not guilty of procuring the suicide by making funeral arrangements from June 4 to 6 2011 in advance of Ms Forde's death.

In defense submissions to the jury, O'Rorke's barrister, Diarmaid McGuinness, senior counsel, told the court that Ms Forde had been denied her constitutional right to travel to Zurich: "There is no offence of making travel arrangements for such purpose," the lawyer said.

The Dublin Circuit Criminal Court has acquitted O'Rorke of the charge of assisting in a suicide.[1] Suicide by gases in England and Wales 2001-2011: evidence of the emergence of new methods of suicide. Paper submitted by Gunnell D, Coope C, Fearn V, Wells C, Hawton K, Kapur N. submitted to the Journal of Affective Disorders.

Appendix 4

Self-chosen deaths by gases in England and Wales

Gunnell et al (2014)[1] have analyzed self-chosen deaths by gassing in England and Wales (population 53 million) using national suicide mortality data. They also analyzed four coroners records of eight people who had used helium. The data below are in a paper they have submitted in 2014 to the Journal of Affective Disorders.

Summary: Gunnell et al have presented evidence that changes in popularity of highly lethal methods can influence their use for suicide. They warn that public health measures are urgently needed to prevent a potential epidemic rise in the use of helium similar to the recent rises in charcoal burning suicides in East Asia. They caution that, "whilst there have been increases in the use of some novel gas methods of suicide in England and Wales, they currently account for a relatively small proportion of total suicides and rises are unlikely to have had an impact on overall suicide rates."

Table: Time trends in gas suicides in England and Wales, 2001-2011 (from Gunnell 2014) *

	2001	2002	2003	2004	2005	2006	2007	2008	2009	2010	2011
Helium	2	3	7	4	2	7	8	22	39	36	53
CO in car exhaust	134	155	120	81	69	58	52	54	46	37	36
CO in Barbecue Charcoal	1	0	1	3	2	6	5	5	4	8	3
CO from unspecified sources **	228	170	188	144	129	100	97	90	67	57	66
Hydrogen sulphide	0	0	0	0	0	0	0	0	0	8	6
Other gases ***	3	9	7	8	3	9	3	6	15	9	10

* The authors give the sum total for each row and column which are omitted here.
** In over 50% of cases the source of carbon monoxide (CO), either charcoal of car exhaust, had not been registered on the death certificate.
*** Nitrogen, argon, carbon dioxide, propane or butane in camping tanks, domestic gas, combustion gas or other gases that are difficult to get hold of.

The data show that the number of helium suicides was relatively constant (2-8 per year) between 2001 and 2007, but it increased significantly from 2008 on. This rise is consistent with reports from San Diego County (population 3 million): from 2 in 2000 and 2001 to 19 in 2011 and 2012. In Australia, (population 23 million) only 79 helium suicides have been registered over a five year period (2005 – 2009) without evidence for an increased use of this route to death.

According to Gunnell individuals who died by helium gas were more likely to be male. A relatively high proportion of these suicides were from managerial and professional occupations. There is evidence of mental health problems in those who have died by helium gas. In the records of four coroners eight helium suicides were identified (mean age 43, range 25-62 years) of whom four were under specialist mental health care around their time of death.

Regarding carbon monoxide Gunnell et al state that in England and Wales the number of suicides with gas exhaust or charcoal are considerably underestimated. Like hydrogen sulfide, carbon monoxide can be a silent killer: they found thirteen cases of accidental deaths by barbecue charcoal (e.g. in campers or tents). Carbon monoxide from charcoal barbecues as a method of suicide has become very popular in Hong Kong and Taiwan. In 2011 it accounted for 13% (Hong Kong) and 24 % (Taiwan) of all suicides.

Suicide with hydrogen sulphide gas was unknown in England and Wales till 2009. Its use got media attention in Japan following a widely reported suicide pact in 2008. In England and Wales eight of the 14 suicides in 2010 and 2011 with this gas, occurred as pairs on the same day, suggesting they were also suicide pacts.. Gunnell identified two cases of accidental deaths with hydrogen sulfide among these 14 deaths.

Comments by Boudewijn Chabot:

The number of self-chosen deaths by helium in the Gunnell study are significant underestimations. Ogden (2001) and others have reported that under repressive legislation the helium equipment was almost always removed prior to reporting death. As helium is not detectable in *standard* post-mortem examinations the cause of death in very ill or very old patients is usually attributed to an underlying illness like 'cardiac arrest'.

Like many other specialists in the field of suicide prevention Gunnell et al ignore the fact that in all western countries some very ill and very old persons long for a physician-assisted death (PAD) before the illness has reached the last terminal weeks. They discuss their wish to hasten their death with the intimate circle of relatives and friends. In jurisdictions where hastening death is legally forbidden, some relatives and friends search the internet for a humane exit, preferably one that is not dangerous for those who want to be present. Some opt for one of the drug methods and order them through the internet. Others choose an inert gas. Right-to-die organizations give information on these routes. Giving information is legal in most countries with the exception (at present) of Australia, New Zealand and some US states.

When the ingredients for a self-chosen humane exit have been collected those who want to be present should learn how to circumvent problems in a post-mortem examination (see appendix 2). It is the legal ban on dignified dying that prompts the intimate circle to cover up what goes on around dying.

Appendix 5

Guideline for the performance of physician-assisted dying in the Netherlands[1]

In 2012, the Royal Dutch Association of Physicians (KNMG) and the Royal Dutch Association of Pharmacists (KNMP) jointly published a new 'Guideline for the performance of physician-assisted dying'. The guideline, which is an update to the 2007 version, discusses an oral and an intravenous method.

The oral method used by doctors

Substances:
- 3 suppositories of metoclopramide 20 mg or 3 tablets of metoclopramide 10 mg;
- 100 ml drink that contains 15 g of pentobarbital. The 15 g of pentobarbital is dissolved in sterile water and 96% alcohol. This mixture has a bitter, soapy taste. Two substances (syrup and aniseed oil) are therefore added to make it less unpleasant.

Preparations for the physician:
- Beforehand, it should be discussed with the patient and possibly with a close relative(s) that if the patient does not die within two hours, the physician will switch to using the intravenous method.
- Start administering metoclopramide one day (twelve hours) in advance. Ideally, metoclopramide should be given at periods of twelve hours, six hours and one hour prior to the performance.

The performance
- Prepare the patient for the fact that it will taste unpleasant.
- Have the patient sit up in bed while drinking. The patient must consume all of the drink.
- The drink should not be taken with a straw, as this carries the danger that the drink will start to have an effect before the whole dose has been taken.

– There are reports of a few cases in which the drink has been administered effectively through a tube into the stomach. Rinse the tube with water immediately, to prevent it from becoming blocked before the barbiturates have reached the stomach or the intestines.
– If the patient vomits up the drink, it is advisable to switch to the intravenous method.
– After the patient has ingested the drink, there is a very high probability that he or she will fall into a deep coma and die.

In the new guideline, the medication for doctor-assisted dying has been increased from the 9 g that used to be recommended to 15 g. This is in order to allow death to occur within a reasonable period of time. Unpublished data from Switzerland indicate that 98% of people who ingest 15 g of pentobarbital die within 30 minutes. If death does not follow after two hours, physicians in the Netherlands are permitted to use the intravenous method.

The intravenous method used by doctors

This method causes death by means of two consecutive injections into a vein. A barbiturate (thiopental 2 g) is first injected to induce a coma, followed by the injection of propofol (150 mg), a chemical substance that inhibits respiration. This method is not discussed in this book because it is impossible for someone to perform this method him- or herself in a self-chosen death.

Appendix 6

Filling, shipping and packaging of nitrogen cylinders

Printed by courtesy of Bob Holub

1. Filling

Airgas is a seller of specialty gases www.airgas.com. Filling the cylinders with nitrogen is not a problem, they will do that if asked to. PurityPlus (purityplusgas.com) lists grades of purity on its website. There are six grades of nitrogen at PurityPlus, all of them are adequate for a self-chosen dignified death. Some tanks are labeled as 'Food grade' which does not refer to the purity of a gas. 'Food grade' refers to the requirement that the distributor maintains a record of the purchaser.

There are small and big cylinders. Would it be possible to get from Airgas a large cylinder and decant it at home into an empty Max Dog Brewing (MDB) cylinder? This would be dangerous, as the difference in pressure between the larger one and the smaller one (2200 psi vs. 1800 psi) might blow out the safety fitting of an MDB steel tank.

2. Shipping

Certification is necessary for anyone directly or indirectly involved with the shipment of hazardous materials ('hazmat' for short) such as compressed nitrogen gas. UPS has an expensive 3 day seminar which satisfies the requirements. There are several firms offering on-line courses. Hazmat School (www.hazmatschool.com) has a DOT (U.S. Department of Transportation) Hazmat class – Course which costs $80 and lasts 5 hours. Their course satisfies the UPS and DOT requirements for training in hazardous materials labeling, shipping and packaging. Another firm has a course which satisfies all ground hazmat carrier requirements including FedEx and UPS compliancetrainingonline.com). It costs $119.95. A course outline can be seen on their web site.

The shippers (e.g. UPS or FedEx) require software prepared paperwork. UPS has an in-house program named WorldShip 2015, there is an 8 page installation guide for the software on the UPS web site. Another firm, Labelmaster, has a very complete program which is used on line. It sells for $349.00 per year has video of required appearance of package for shipping.

3. Packaging

The course outline for hazmat training has several headings on this subject. UPS documentation refers to DOT requirements and says that cylinders must be 'overcartoned'. For FedEx a good solid cardboard box with stuffing to keep the cylinder from moving about would be adequate.

Once the cylinder is properly packaged and labeled, and the shipping documentation software prepared either UPS or FedEx will take it from there. There is a UPS surcharge for each dangerous goods shipment of $28.50 over its normal cost of shipping. Using the weight and measurements of a MDB tank and box, for shipment from California to Florida FedEx will charge (assuming there is an account plus a contract in place (both UPS & FedEx require these]: $176.56 priority air, $161.36 standard overnight air, $106.67 2 day, $79.45 3 day.

The contract required by UPS and by FedEx would probably refer to liability. It is likely that all liability for problems of any nature flow to the shipper, not the carrier (UPS or FedEx).

Notes

2. Edinboro 1997; Kramer 1998
3. Admiraal 2006
4. Buckley 1996
5. Docker 2013 pp 298 and 307

Intermezzo: self-chosen death by helium gas
1. Reerink 2013
2. www.dyingathome.nl
3. Chabot 2010

Chapter 3
1. Auwärter 2004
2. Ogden 2010a
3. Poklis 2003, Schön 2007, Schaff 2012
4. Ogden 2002, Ogden 2010a
5. Ogden ch. 8 in Admiraal, Chabot and Ogden 2006
6. Horikx 2000
7. Chabot 2012
8. Gunnell 2015
9. Ogden 2001
10. Poklis 2003, Schön 2007
11. Ogden 2010b
12. Bosshard 2003, Bosshard 2008
13. Putz 2012

Chapter 4
1. Seale 2004
2. Economist Intelligence Unit 2014
3. Chabot 2007, Chabot 2009
4. Nitschke 2006
5. Horikx 2000
6. Syme 2008
7. Humphry 2010 p 151
8. Ogden 1994, Jamison 1996, Chabot 1996, Magnusson 2002
9. Admiraal 2006
10. Admiraal 2006. Docker 2013
11. Pridonoff 1996

Preface
1. Griffiths 1995, Klotzko 1995, Sneiderman 1996
2. Achterhuis 1995, Hendin 1997. Wijsbek 2010
3. Cassell 1991
4. Chabot 2009, Chabot 2012
5. Chabot 2010, Chabot 2014
6. Badham 2009
7. Chabot 2007, Van der Heide 2012

Introduction
1. The Economist 2015
2. Seale 2004, Kellehear 2007
3. Terman 2007
4. Chabot 2014
5. Cavendish 2014
6. Quill 1997, Ganzini 2004, Harvath 2006
7. KNMG 2015
8. Park 2014
9. EHRC 2011, EHRC 2012
10. Heringa 2015
11. Belfast Telegraph 24-4-2015 (see Appendix 3)
12. Chabot 2009

Chapter 1
1. Filene 1998
2. Sullivan 1998
3. Hagens 2014

Intermezzo: self-chosen death by pentobarbital
1. Widow Nel in film on dvd, Chabot 2013

Chapter 2
1. http://www.oregon.gov/DHS/ph/pas/ar-index.shtml, retrieved May 23, 2006

23. Smith 1995, Dawson 2004
24. Levy 1971
25. Smith 1995, Wolff 2002
26. Booij 2008
27. Nelson 2006

Appendix 1
1. Terman 2007 gives a comprehensive discussion

Appendix 4
1. Gunnell 2015

Appendix 5
1. KNMG & KNMP guideline 2012

12. Docker 2013
13. Nitschke 2006 online edition
14. NuTech 2015 expert meeting San Francisco
15. Gunnell 2015
16. Neal Nicol personal communication 2014
17. Nitschke 2006
18. Gerald Metz personal communication
19. Docker 2013
20. Case of Albert Heringa 2015
21. Ogden personal communication 2013
22. Ogden 2010b

References

The titles of Dutch references have been translated into English

Achterhuis H. 1994 *Als de dood voor het leven* (in Dutch: when death becomes more important than life). Van Ooschot

Auwärter, V. Pragst, F. & Strauch, H. (2004). Analytical investigations in a death case by suffocation in an argon atmosphere. *Forensic Science International*, 143, 169-175.

Badham, P. (2009). *Is there a Christian Case for Assisted Dying? Voluntary euthanasia reassessed*. London: Society for Promoting Christian Knowledge.

Belfast Telegraph on Dublin Circuit Criminal Court 24 April 2015. The case of Gail O'Rorke.

Booij L.H.D.J. and Vree T.B. 2002 Neuromuscular transmission and neurotoxins. (in Dutch). Ned. T. voor anesthesiology 15 Dec. 2002, 128-137.

Bosshard, G., Ulrich, E., & Bär, W. (2003). 748 cases of suicide assisted by a Swiss right to die organization. *Swiss Medical Weekly*, 133, 310-317

Bosshard G. (2008). Switzerland. In: J. Griffiths, H. Weyers, M. Adams (2008). *Euthanasia and Law in Europe*. Oxford: Hart

Buckley, N.A., Smith, A.J., Dosen, P., & O'Connell, D.L. (1996). Effects of catecholamines and diazepam in chloroquine poisoning in barbiturate anaesthetized rats. *Human & experimental toxicology*, 15, 909-914.

Cavendish C. in: The Sunday Times 20-04-2014.

Cassell, E.J. 1991 *The Nature of Suffering and the Goal of Medicine*. Oxford University Press.

Chabot B. 2007 (in Dutch) *Auto-euthanasie*. Verborgen stervenswegen in gesprek met naasten. [*A humane self-chosen death. Hidden dying trajectories in conversation with proxies*]. PhD thesis University of Amsterdam. Bert Bakker Publishers.

Chabot B, A. Goedhart. 2009. A survey of self-directed dying attended by proxies in the Dutch population. *Social Science & Medicine* 68:1745-1751.

Chabot B., S. Braam. 2010 (in Dutch) *Uitweg. Een waardig levenseinde in eigen hand*. [A Way to Die. Handbook. A dignified and self-directed death]. Amsterdam: Nijgh & Van Ditmar

Chabot B. 2012 Hastening Death Through Voluntary Cessation of Eating and Drinking: A Survey. in: Youngner SJ, Kimsma G.K. (eds). *Physician-Assisted Death in Perspective. Assessing the Dutch Experience*. New York: Cambridge University Press.

Chabot B. 2014. *Taking Control of Your Death by Stopping Eating and Drinking*. Amsterdam: Foundation Dignified Dying.

Chabot B. 2012 Film English spoken on dvd. *The helium method. Dying at home with helium*. Amazon.com or www.dyingathome.nl

Chabot B. 2013. Film English spoken on dvd:. *Eyewitnesses. Personal Narratives of self-chosen and humane deaths.* Amsterdam: Foundation Dignified Dying.

Chabot, B.E., K. Gill 1996 (in Dutch) De plastic zak methode. [The plastic bag method for a self-chosen death]. *Tijdschrift voor Huisartsgeneeskunde*, 13, nr 3.

Chabot B. & C. Walther. 2013 (in German). *Ausweg am Lebensende. Sterbefasten — Selbstbestimmtes Sterben durch freiwilligen Verzicht auf Essen und Trinken.* [The way to die by voluntary stopping eating and drinking]. 3. Auflage München: Reinhardt Verlag.

Clemessy J.L., Taboulet, P., Hoffman, J.R., Hantson, P., Barriot, P., Bismuth, C., & Baud, F.J. (1996). Treatment of acute chloroquine poisoning: A 5-year experience. *Critical Care Medicine*, 24, 1189-1195.

Clemessy J.L., Angel, G., Borron, S.W., Ndiaye, M., Le Brun, F., Julien, H., Galliot, M., Vicaut, E., & Baud, F.J. (1996). Therapeutic trial of diazepam versus placebo in acute chloroquine intoxications of moderate gravity. *Intensive Care Medicine*, 22, 1400-1405.

Coté J. (2012) *In Search of a Gentle Death. The Fight for Your Right to Die with Dignity.* Corinthian Books. ISBN-13: 978 19291 75369

Court Case of Albert Heringa. Court of Appeal Arnhem / Leeuwarden 06-05-2015

Dawson, A.H. (2004). Cyclic Antidepressant Drugs. In: R.C. Dart (ed.):   *Medical Toxicology*, 3d edition. Philadelphia (USA): Lippincott Williams & Wilkins: 834-843.

Demaziere, J., Saissy, J.M., Vitris, M., Seck, M., Ndiaye, M., Gaye, M., & Marcoux, M. (1992). Effects of diazepam on mortality from acute chloroquine poisoning. *Annales Françaises d'anesthèsie et de Rèanimation* 11, 164-167.

Docker C. 2013. Five Last Acts. The Exit Path. The arts and science of rational suicide in the face of unbearable, unrelievable suffering. Edinburgh, 3d ed.

DrugDex. Micromedex Healthcare Series, Healthcare business of Thomson Reuters. (via internet subscription).

Eckholm E. 'Aid in Dying' movement takes hold in some states. *The New York Times* Feb 7, 2014. [Oregon, Washington, Montana, Vermont, New Mexico]

Economist Intelligence Unit (2014). The Quality of Death: Ranking End-of-Life Care across the world. EIC.com 12-05-2014

EHRC *European Human Rights Cases*, Haas v. Switzerland, 20/1/2011. nr 31322/07. Commentary Isra Black in *Medical Law Review* 20 (2012), 157-176.

EHRC *European Human Rights Casses*, Koch v. Germany 19/07/2012, nr 497/09.

Edinboro, L.E., Poklis, A., Trautman, D. Lowry, S., Backer, R. & Harvey, C.M. (1997). Fatal fentanyl intoxication following excessive transdermal application. *Journal of Forensic Science*, 42, 741-743

Filene P.G. (1998). *In the Arms of Others. A Cultural History of the Right-to-Die in America*. Chicago: Ivan R. Dee

Gallagher, K.E., Smith, D.M., & Mellen, P.F. (2003). Suicidal Asphyxiation by Using Pure Helium Gas: Case Report, Review and Discussion of the Influence of the Internet. *The American Journal of Forensic Medicine and Pathology*, 24, 361-363.

Ganzini L, Goy E.R., Miller L.L. et al 2003. Nurses Experiences with Hospice Patients who Refuse Food and Fluids to Hasten Death. New England Journal of Medicine; 349:359-365.

Griffiths J. (1995) Assisted Suicide in the Netherlands: The Chabot Case. *The Modern Law Review*, Vol. 58, No. 2 (Mar., 1995), pp. 232-248

Griffiths J., Weyers H., Adams M. (2008) *Euthanasia and Law in Europe*. Oxford and Portland (Oregon) Hart Publishing.

Gunnell D., Coope C., Fearn V., Wells C. et al. 2015 Suicide by gases in England and Wales 2001-2011: evidence of the emergence of new methods of suicide. J Affect Disord. Jan 1;170:190-5.

Hagens M, Pasman R.H., Onwuteaka-Philipsen B.D. 2014 Cross-sectional research into counseling for non-physician assisted suicide: who asks for it and what happens? BMC Heath Services Research, 14:455 www.biomedcentral.com

Hartogh G. den 2012 The Regulation of Euthanasia: How Successful is the Dutch System? In: Youngner S.J., Kimsma G.K. (eds). *Physician-Assisted Death in Perspective. Assessing the Dutch Experience*. New York: Cambridge University Press.

Hartogh, G. den 2014 (in Dutch). Hulp bij zelfdoding door intimi. Een grondrechtconforme uitleg van artikel 294 Sr. [Assisted suicide by intimates. An interpretation of art 294 in Dutch penal law that is consistent with the Constitution]. *Nederlands Juristenblad*, 89 (2014) 24, 1596-1663.

Harvath T.A., Miller L.L., Smith K.A. et al. 2006 Dilemmas encountered by hospice workers when patients wish to hasten death. *Journal of Hospice and Palliative Nursing*;8:200-209

Heide, A van der, Deliens, L., Faisst, K. et al (2003). End-of-life decision-making in six European countries: a descriptive study. *The Lancet*, 362, 345-350.

Heide A van der, Brinkman-Stoppelenburg A, Delden J.J.M. van, Onwuteaka-Philipsen, B.D. 2012 (in Dutch) *Sterfgevallenonderzoek 2010. Euthanasie en andere medische beslissingen rond het levenseinde*. [Investigation into deaths in 2010. Physician-assisted dying and other medical decisions relating to the end of life]. Den Haag: ZonMw; 2012.

Hendin, H. 1997. *Seduced by Death*. New York: W.W. Norton & Company: 68

Heringa A., 2015. Verdict of the Court of Appeal Arnhem-Leeuwarden, 13-05-2015. Case nr 21-008160-13

Hooff, A van. 1990 (in Dutch). *Zelfdoding in de antieke wereld*. Suicide in the ancient world] Nijmegen: SUN.

Horikx A. & Admiraal, P.V. 2000 (in Dutch). Toepassing van euthanatica; erva-ringen van artsen bij 227 patiënten, 1998-2000 [Application of euthanatics; physicians' experiences with 227 patients, 1998-2000]. *Ned Tijdschr Geneeskd*,

144, 2497-2450.

Howard M.O., Hall M.T., Edwards J.D., et al. Suicide by Asphyxiation due to Helium Inhalation. *Am J Forenxic Med Pathol*, 32 (1), 61-70.

Humphry D. (2002) *Final Exit. The practicalities of Self-deliverance and Assisted Suicide for the Dying*. New York: Random House third edition revised and updated. 2010 Delta Trade Paperback Edition.

Jamison S. 1996. When Drugs Fail: Assisted Deaths and Not-So-Lethal drugs. In: Battin, M. P. & Lipman, A. G. (eds). *Drug use in Assisted Suicide and Euthanasia* (pp. 223-243). New York: The Pharmaceutical Products Press.

Kellehear A. 2007 *A Social History of Dying*. Cambridge: Cambridge University Press.

Klotzko, A.J. 1995. Arlene Judith Klotko and Dr. Boudewijn Chabot Discuss Assisted Suicide in the Absence of Somatic Illness. *Cambridge Quaterly of Healthcare Ethics*, 4, 239-249.

KNMG (Royal Dutch Medical Association). 2011 Position Paper: the Role of the physician in the voluntary termination of life. Utrecht www.knmg.nl

KNMG, 2015. Caring for people who consciously choose not to eat and drink so as to hasten the end of life. Utrecht www.knmg.nl

KNMG & knmp 2012 (in Dutch) Richtlijn uitvoering euthanasie en hulp bij zelfdoding. [Guideline for performing physician-assisted dying, revised edition] Utrecht. www.knmg.nl

Kramer, C. & Tawney, M. (1998). A fatal dose of transdermally administered fentanyl. *Journal of the American Osteopathic Association*, 98, 385-386.

Levy, A.H. 1971 (in Dutch) *Digitalin intoxication*. Dissertation, Leiden. University

Lewis P. England and Wales. In: J. Griffiths, H. Weyers, M. Adams (2008). *Euthanasia and Law in Europe*. Oxford: Hart Publishers.

Magnusson R.S. 2002 *Angels of Death. Exploring the Euthanasia Underground*. New Haven and London: Yale University Press.

Martindale. The complete Drug Reference. Sweetman S.C., Wolters Kluwer, Pharmaceutical Press. ISBN-13: 9780853696872.

Nelson L. 2006. Acute Cyanide Toxicity: Mechanisms and Manifestations. *J of Emerg Nursing* 32:4S, pp S8-S11, Aug 2006

Nitschke Ph., Stewart F. 2006. *The Peaceful Pill Handbook*. Online edition. Exit International US.

Ogden, R.D. 1994. *Euthanasia, assisted suicide & AIDS*. New Westminster: Peroglyphics Publishing.

Ogden, R.D. 2001. Non-physician-assisted suicide: The technological imperative of the deathing counterculture. *Death Studies*, 25, 387-401.

Ogden, R.D. & Wooten, R. H. (2002). Asphyxial suicide with helium and a plastic bag. The American Journal of Forensic Medicine and Pathology, 23, 234-237.

Ogden R.D. (2010a). Observations of Two Suicides by Helium Inhalation in a Prefilled Environment. *The American Journal of Forensic Medicine and Pathology*, 31: 156-161.

Ogden R.D. (2010b). Assisted Suicide by Oxygen Deprivation with Helium at a Swiss Right-to-Die Organization. *Journal of Medical Ethics*, 36:174-179.

Oregon Department of Human Services, Office of Disease Prevention and Epidemiology, 2006, March 9. Eighth annual report on Oregon's Death with Dignity Act. Oregon: DHS., from http://www.oregon.gov/DHS/ph/pas/ar-index.shtml May 23, 2006.

Poklis J.L., Garside D., Winecker R.E. et al. 2005. A qualitative method for the detection of helium in postmortem blood and tissues. In: *Proceeding from the Society of Forensic toxicologists*, Oct. 17-11, 2005 Nasville TN

Pridonoff J. 1994. In: Hemlock Time Lines

Putz, W., Steldinger B. 2012. *Patientenrechte am Ende des Lebens: Vorsorgevollmacht, Patientenverfügung, selbstbestimmtes Sterben.* 4. Aufl. DTV, München.

Reerink A. 2013 (in Dutch) Sterven met heliumgas: 'Dit is humaner dan voor de trein springen' [Dying with helium gas is more humane than jumping in front of a train']. in: NRC *Handelsblad* Newspaper 15-09-2012.

Richards, N. 2012. The fight-to-die: older people and death activism. *International Journal of Ageing and Later Life*;7:7-32

Schaff J. F., Karas R.P., Marinetti L. (2012) A Gas-Chromatography-Thermal Conductivity Detection Method for Helium Detection in Postmortem Blood and Tissue Specimens. *Journal of Analytical Toxicology* 2012;36:112-115.

Schön C.A.,Ketterer T. (2007) Asphyxial Suicide by Inhalation of Helium Inside a Plastic Bag. *The American Journal of Forensic Medicine and Pathology* 2007;28:364-367.

Seale C., Geest S. van der 2004. 'Good and Bad Death': Introduction to Special Issue of *SocialScience and Medicine*. 58:883-995

Smith, C.K. (1995). Street drugs. In Smith, C.K., Docker, C.G. & Hofsess, J. (Eds.), Beyond Final Exit. Victoria: Right to Die Society of Canada.

Smith, C.K. (1995). Tricyclic antidepressants: A new look. In C.K. Smith, Docker, C.G., Hofsess, J. & Dunn, B. (Eds.), Beyond Final Exit (pp. 29-39). Victoria: Right to Die Society of Canada.

Sneiderman, B, M. Verhoef. 1996. Patient Autonomy and the Defence of Medical Necessity: Five Dutch Euthanasia Cases. *Alta Law Rev* 34: 374-415.

Sullivan M.D., Ganzini L., Youngner S.J. 1998. Should Psychiatrists serve as Gatekeepers for Physician-Assisted suicide? *Hastings Center Report* 28. no 4: 24-31.

Syme R. 2008. *A good Death. An Argument for Voluntary Euthanasia.* Melbourne

University Press.

Terman S.A. 2006. *The Best Way to Say Goodbye. A Legal Peaceful Choice at the End of Life.* Carlsbad: Life Transitions Publications.

Vink, T. 2013. (in Dutch) *Zelfeuthanasie. Een zelfbezorgde goede dood onder eigen regie.* [Self-euthanasia. A self-directed and good death performed by oneself] Damon.

Wolff, K. 2002. Characterization of methadone overdose: clinical considerations and the scientific evidence. *Therapeutic Drug Monitoring,* 24, 457-470.

Wozz 2006. Admiraal P., Chabot B., Ogden R.D. Glerum J. *Guide to a humane self-chosen death.* Amsterdam: Wozz Foundation.

Wijsbek H. 2010. 'To thine own self be true': On the loss of integrity as a kind of suffering. *Bioethics* 26(1);1-7.

Index

Made in United States
Troutdale, OR
08/09/2024

21864853R50073